Market Power Politics

Market Power Politics

War, Institutions, and Strategic
Delay in World Politics

STEPHEN E. GENT

AND

MARK J.C. CRESCENZI

OXFORD
UNIVERSITY PRESS

OXFORD
UNIVERSITY PRESS

Oxford University Press is a department of the University of Oxford. It furthers
the University's objective of excellence in research, scholarship, and education
by publishing worldwide. Oxford is a registered trade mark of Oxford University
Press in the UK and certain other countries.

Published in the United States of America by Oxford University Press
198 Madison Avenue, New York, NY 10016, United States of America.

© Oxford University Press 2021

Library of Congress Cataloging-in-Publication Data
Names: Gent, Stephen E., 1976– author. | Crescenzi, Mark J.C., 1970– author.
Title: Market power politics : war, institutions, and strategic delay in
world politics / Stephen E. Gent and Mark J.C. Crescenzi.
Description: New York, NY : Oxford University Press, [2021] |
Includes bibliographical references and index.
Identifiers: LCCN 2020034295 (print) | LCCN 2020034296 (ebook) |
ISBN 9780197529812 (paperback) | ISBN 9780197529805 (hardback) |
ISBN 9780197529836 (epub)
Subjects: LCSH: International economic relations. | Natural resources. |
Globalization—Economic aspects. | War—Economic aspects.
Classification: LCC HF1359.G4688 2021 (print) | LCC HF1359 (ebook) |
DDC 355.02/73—dc23
LC record available at https://lccn.loc.gov/2020034295
LC ebook record available at https://lccn.loc.gov/2020034296

DOI: 10.1093/oso/9780197529805.001.0001

1 3 5 7 9 8 6 4 2

Paperback printed by Marquis, Canada
Hardback printed by Bridgeport National Bindery, Inc., United States of America

For Nick and Anita

Contents

Figures

Tables

Preface

IN ITS BROADEST strokes, *Market Power Politics* is a book about politics, economics, and law in the global arena. The politics we cover here deal with peace and conflict between countries (we will call them *states* in the book). With respect to economics, we are interested primarily in trade and investment that crosses international borders. Lastly, we care here about the use of international laws and institutions that states have created to manage either the first (politics) or the second (economics) component, or both. Sometimes these three things fit together efficiently, making it easier for firms and their governments to exchange goods, services, and capital in ways that tie the world together and make it more peaceful. When that happens, the three components start to work together as a team. Legal institutions make it easier to trade goods and services or invest capital, for example, and increased wealth from trade may minimize the desire to interrupt that trade with political violence.

Other times, however, at least one of the pieces is out of alignment with the others. When that happens, the same connectivity that brings these three dimensions of human interaction together as a team can lead to difficult or even disastrous consequences. When institutions cannot offer a legal solution to disputes, states will seek their own solutions, often at the expense of others. When those disputes are motivated by *market power*, or the ability to influence prices, supply, or demand within markets—particularly in markets that are important to governments and their leaders—then the logic of minimizing political violence in the interest of wealth can flip on its head. This is a book about the consequences of that flip. The consequences can be quite dramatic, resulting in the use of military force and even the outbreak of war. But often the consequences of market power fall in between the two extremes of prosperous peace and the catastrophic consequences of war. Instead, states sometimes opt out of cooperation with one another; but rather than jumping right into a fight, they *delay* a resolution to their dispute. They may even push one another around while they are doing it, engaging in bellicose talk and small territorial or maritime grabs without ever boiling over into full-scale war.

The view we take in this book is that even in these situations where the synergy of politics, economics, and law fails to materialize and market power opportunities trigger conflict, the three components still work together in important ways. By studying their interaction even when things aren't going smoothly, we believe we can identify why some states boil over into war while others seem to simmer incessantly. We illustrate our intuition by applying our theory of market power politics to three of the most important places and moments in the last fifty years: the Persian Gulf War in the 1990s, Russian conflict with its neighbors such as Georgia and Ukraine in the previous decade, and China's interactions with its neighbors in the East and South China Seas over the last thirty years.

This already sounds complicated, so let us pause here to reassure the reader that we have designed this book to be widely accessible. Our hope is that anyone with an interest in world politics can read the book, and we have removed as much of the technical jargon as we could in pursuit of this goal. Non-academic readers may want to skim through Chapter 4, however, as we do spend a little time talking about how we chose these cases and why that matters. Even so, anyone who has taken an introductory course in international relations should have the tools to read the entire book.

In some ways, this book represents the partnership of its two authors. Stephen Gent has long held an interest in the interactions between international institutions and peace. His expertise in the goals and structures of these institutions helped us see the cause of their failure to resolve the disputes that we examine in the second half of the book. Mark Crescenzi has studied the link between economics and conflict in world politics for a quarter of a century. When he wrote his first book on economic interdependence and conflict, and how interdependence can sometimes keep states from fighting, this flip side of the relationship nagged in the back of his mind. Both of us found that we were unable to complete our thoughts on these matters on our own. But while sitting on a bench on the balcony of the UNC Student Bookstore, we hatched a plan to work together. This book is the product of those efforts.

Along the way we have had a tremendous amount of help. We presented nascent pieces of the research at academic conferences and meetings, gathering feedback from scholars such as Kyle Beardsley, Stephen Chaudoin, David Cunningham, Scott Gartner, Paul Hensel, Kelly Kadera, Pat McDonald, Sara Mitchell, Desiree Nilsson, Jack Paine, Michael Reese, Aisling Winston, and especially Krista Wiegand, whose work was also foundational for our own. At UNC we sought feedback and moral support from Cameron Ballard-Rosa, Navin Bapat, Graeme Robertson, and Tricia Sullivan. We simply could not ask for more supportive colleagues. Our students were amazing, too, and we would like to thank Michelle Corea, Tyler Ditmore, Bailee Donahue, Derek Galyon, Dan Gustafson, Austin Hahn, Rebecca Kalmbach, Justin Kranis, Emily Rose Mitchell, Lauren Morris, Eric Parajon, Michael Purello, Steven Saroka, Maya Schroder, Stephanie Shady, Zach Simon, Michelle Smoler, Kai Stern, Anna Sturkey, and Rob Williams for all of their research assistance. Thanks also to David McBride, Holly Mitchell, Gopinath Anbalagan, and the production team at Oxford University Press for guidance and support throughout the publication process. We are also grateful to Don Larson and the team at Mapping Specialists, who created the original maps for the book.

Finally, Stephen would like to thank Ed and Joy Gent for providing unwavering support and encouragement over the years. He would also like to thank Michael Cain and J. P. Singh, who introduced him to the study of political science and international relations as an undergraduate at Ole Miss. Their classes undoubtedly laid the seed for many of the questions tackled in this book. Mark would like to thank Jim Crescenzi for patiently serving as an early reader, but also for a lifetime of inspiration, friendship, and encouragement. And, of course, we never would have made it to the end of the project without extraordinary support from Nick Siedentop and Anita Crescenzi. We couldn't imagine better partners as we sorted this out, and we are so thankful that they put up with us.

Abbreviations

BIS	Bank of International Settlements
DPR	Donetsk People's Republic
DSB	Dispute Settlement Body (of the World Trade Organization)
EEZ	exclusive economic zone
EU	European Union
FDI	foreign direct investment
GATT	General Agreement on Tariffs and Trade
ICJ	International Court of Justice
INOC	Iraq National Oil Company
ISA	International Seabed Authority
JCG	Japanese Coast Guard
KOC	Kuwait Oil Company
LNG	liquified natural gas
LPR	Luhansk People's Republic
NATO	North Atlantic Treaty Organization
NEE	normal economic exchange
OPEC	Organization of the Petroleum Exporting Countries
PCA	Permanent Court of Arbitration
PRC	People's Republic of China
REE	rare earth element
TANAP	Trans-Anatolian Natural Gas Pipeline
TAP	Trans-Adriatic Pipeline
tcm	thousand cubic meters
UAE	United Arab Emirates
UN	United Nations
UNCLOS	United Nations Convention on the Law of the Sea
WTO	World Trade Organization

I

Introduction

A MIDNIGHT INVASION of Iraqi troops into Kuwait in 1990 set off a chain of events that have reshaped politics in the Middle East with devastating and deadly consequences. Nearly twenty-five years later, Russia annexed the Ukrainian territory of Crimea and pushed the envelope of using military might without triggering war. Meanwhile, in the South China Sea, China has made bold moves to expand its territory island by island, triggering legal disputes from its neighbors, and yet the fear of militarized violence seems relatively low compared to Russia or Iraq. What do these crucial geopolitical moments in history have in common, besides their importance in the lives and prosperity of dozens of countries and millions of people? Could it be that all three cases are being driven by a common political-economic process? If so, what explains the variance in their outcomes? How do we know when a situation like these will become violent?

Answering these questions is the goal of this book. Here we develop a theory of market power politics that helps us understand why states abandon the post–World War II institutions that help states navigate territorial disputes such as these. In this analysis we identify three potential outcomes when market power politics dominate: war, cooperation, and strategic delay. Understanding the context of political institutions and economic ties that surround market power opportunities is key to knowing which outcome is likely to result when states pursue a strategy of dominating a market. In an era where we have come to expect institutions, trade, and investment to knit governments together and prevent violence, in this book we identify one source of motivation that unravels the fabric of peace and increases the risk of war.

Strategic Delay in the "Gray Zone"

The ambition of states to expand their geographic reach can create or exacerbate disputes over the control of territory or maritime areas. We refer to such disagreements as international *property rights disputes* because they involve competing claims over the sovereign control of resources. States have several strategies to choose from as they try to resolve these property rights disputes in their favor. Some, like Iraq in 1990, choose to press their claims on the battlefield. In fact, disputes over territory have historically been one of the most predominant causes of international war.[1] Alternatively, states may try to avoid a costly war and instead work toward a peaceful resolution of their dispute. Through bilateral negotiations or with the assistance of international institutions, states can reach agreements over how to divide disputed territory or maritime areas.

However, sometimes states choose to go down neither of these paths. Instead, they opt to pursue a policy of strategic delay. *Strategic delay* is the purposeful postponement of a violent or nonviolent settlement of a dispute with the hope of achieving a more preferable outcome in the future. The use of strategic delay is not uncommon in property rights disputes, which are often long lasting.[2] In some cases, these delay strategies are largely passive in nature. States may simply want to maintain the status quo because they do not expect that they would achieve a more favorable outcome through military force or negotiation.[3] At other times, though, we see states taking a more proactive approach and using strategic delay as a way to further their expansionist goals. By delaying, these states have the opportunity to pursue so-called gray zone tactics that allow them to gradually shift the strategic environment in their favor over time.

Gray zone tactics lie somewhere in the "gray zone" between diplomacy and war. They do not involve the overt use of lethal force by a country's military, so they fall short of what we would call strategies of war. On the other hand, they also fall outside the bounds of the behavior traditionally acceptable within international diplomacy. These gray zone activities often take the form of *salami tactics*. Like consuming an entire salami by cutting a series of small slices, salami tactics involve a country taking small steps that accumulate and achieve a larger goal over time.[4] In property rights disputes, such a gradual approach can be very effective, as each of the individual steps is less likely to provoke an armed response

---

1. Holsti 1991; Vasquez 1993.

2. Wiegand 2011.

3. Fravel 2008; Huth, Croco, and Appel 2011.

4. Schelling 1966.

than more aggressive actions. In this way, gray zone tactics allow states to press their territorial and maritime claims while also hoping to avoid major armed conflict. To get a sense of how this combination of strategic delay and gray zone tactics plays out empirically, let us take a look at some of the expansionist activities pursued by China and Russia in recent years.

Mischief in the South China Sea

Consider, for example, the case of Mischief Reef and the ongoing dispute between China and the Philippines. Mischief Reef (Figure 1.1) is an atoll in the South China Sea, roughly 135 miles west of the Philippines. Part of the Spratly Islands, it consists of a narrow coral reef that surrounds a large lagoon. Up until a few years ago, Mischief Reef was what hydrographers and international lawyers would call a low-tide elevation, as it only rose above the level of sea during low tide. Needless to say, for most of its history, Mischief Reef has been home only to marine life and was primarily visited by fishermen. All of that began to change in the mid-1990s, however, and today the atoll lies at the center of one of the world's

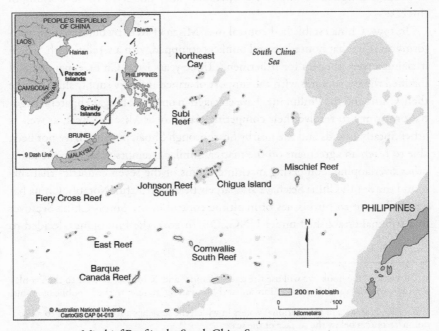

FIGURE 1.1. Mischief Reef in the South China Sea.
Copyright: CartoGIS Services, College of Asia and the Pacific, The Australian National University, Creative Commons License, CC BY-SA 4.0.

most contentious maritime disputes. What was once a tiny slice of earth that only poked its head above water at low tide has now been launched out of obscurity and is a serious source of international tension in the Asia-Pacific region.

While four countries—China, the Philippines, Taiwan, and Vietnam—make claim to Mischief Reef, China and the Philippines have taken the most active steps to assert control over the atoll.[5] China's claim to Mischief Reef is historical. Based upon mid-twentieth-century maps that include the so-called Nine-Dash Line, China contends that it has jurisdiction over almost all of the South China Sea. Thus, despite the fact that Mischief Reef lies over 600 miles from China's Hainan Island and about 800 miles from the Chinese mainland, China claims that the reef is part of its territory. The Philippines, on the other hand, claims that Mischief Reef lies within its own exclusive economic zone (EEZ), a maritime area in which it has jurisdiction over the exploration and exploitation of marine resources.[6] The Philippine government's claim is based upon its interpretation of the rules for drawing international maritime boundaries laid out in the United Nations Convention on the Law of the Sea (UNCLOS). Both China and the Philippines are parties to UNCLOS, which prioritizes the type of rules-based claim made by the Philippines over historical claims. However, this international institution designed to improve cooperation does not seem to be working in this case.

In 1995, China established control over Mischief Reef by building four platforms on stilts that housed several bunkers equipped with a satellite dish, which it claimed were shelters for fishermen.[7] Three years later, China expanded and fortified these structures with the support of armed military supply ships.[8] Since then, China and the Philippines have engaged in on-and-off-again bilateral negotiations to try to resolve their competing claims over Mischief Reef, as well as other Spratly Islands and the nearby Scarborough Shoal, but they have not been able to reach an agreement on the matter. Similar attempts to resolve the many other overlapping, competing maritime claims of the seven countries that surround the South China Sea have also largely been unfruitful. Notably, China has been reluctant to turn issues of maritime control in the South China Sea over to a tribunal established under UNCLOS. In 2013, the Philippines decided to

---

5. Throughout the book, we will use the terms "China" and "Chinese" to refer to the People's Republic of China and the terms "Taiwan" and "Taiwanese" to refer to the Republic of China.

6. An EEZ is an area in international waters in which the state has a sovereign right to all economic resources below the surface of the sea.

7. Shenon 1995; Storey 1999.

8. Zha and Valencia 2001.

take matters into its own hands and notified China that it was going to unilaterally bring its case to the Permanent Court of Arbitration (PCA). China quickly refused to participate. While the Philippine government was eager to use legal institutions to resolve this dispute, China seemed equally eager not to do so.

While the Philippines turned to the courts, China began to take more aggressive steps to expand its presence in the South China Sea. A significant part of China's strategy involved the construction of artificial islands on its occupied features in the Spratly Islands, including Mischief Reef. In January 2015, China began dredging sand and pumping it on top of the coral at Mischief Reef while an amphibious warship patrolled the entrance to the reef's lagoon.[9] By June of that year, the Asia Maritime Transparency Initiative estimated that over two square miles of land had been already been reclaimed.[10] Since then, Chinese development on Mischief Reef has moved apace, and it continues to construct more permanent facilities on the island. A three-kilometer-long runway was completed in 2016, along with several hangars for large aircraft. Additional construction projects on the island have included underground storage for ammunition, communications towers, and structures to house defensive weapons systems.[11]

An adverse ruling by the PCA in June 2016 hardly provided a speed bump to China's expansionist plans on Mischief Reef. The Court ruled that the reef was a low-tide elevation before China's reclamation efforts and thus does not generate any entitlement to a territorial sea. Moreover, the Court found that Mischief Reef lies within the Philippines' EEZ. China, however, refused to accept the court's ruling and continued to claim sovereignty over much of the South China Sea. Rather than taking a step back, China pushed forward with its efforts to militarize the Spratly Islands, including Mischief Reef. By mid-2018, the Chinese military had quietly installed communications jamming equipment, anti-aircraft guns, anti-ship cruise missiles, and surface-to-air missile systems on its outposts in the Spratly Islands.[12]

The pattern of activities on Mischief Reef mirrors similar Chinese projects on other islands in the Spratly archipelago, including Fiery Cross Reef and Subi Reef. Whether or not it had initially aimed to just build structures to provide shelter for fishermen in the mid-1990s, China has clearly moved toward a slow but steady militarization of its claimed territory in the South China Sea. Rather

---

9. Sanger and Gladstone 2015.

10. Asia Maritime Transparency Initiative 2015.

11. Asia Maritime Transparency Initiative 2018.

12. Macias 2018.

than accepting a legal settlement or turning to military force, China has preferred a policy of strategic delay. By delaying, China has been able to pursue a campaign of gray zone activities to expand its presence in the South China Sea. In the case of Mischief Reef, China's strategy provides a textbook example of salami tactics. China first installed platforms on the reef, followed by the construction of an artificial island that could house larger structures and an airstrip. It is now moving forward with the installation of military weapons on the island. These steps have allowed China to gradually consolidate its control of Mischief Reef and other islands the South China Sea without sparking a major armed conflict.

Russian Expansion: A Land-Based Analogue

While China has been building islands to expand its reach into the South China Sea, Russia has pursued territorial ambitions of its own in the Black Sea region. South Ossetia is a landlocked breakaway region of north-central Georgia that borders Russia. It has little in the way of natural resources, and its citizens largely rely upon subsistence farming. In the wake of the breakup of the Soviet Union, South Ossetia declared independence from Georgia, sparking the first of a series of civil wars that Georgia suffered in the early 1990s. Since then, South Ossetia has often been described as being a "frozen conflict," and the region has largely remained outside the effective control of the Georgian government, relying upon the support of Russia for survival. As part of the civil war peace settlement in 1992, Russia deployed a peace-keeping force to the region that has never left. In the mid-2000s, Russia gradually increased its military capabilities in South Ossetia, and during a brief five-day war in August 2008, Russian and Ossetian troops pushed out any remaining Georgian presence in the region. Later that month, Russia officially recognized South Ossetia's independence, but very few other countries followed suit.

Since the 2008 conflict, Russia has continued to expand its influence in South Ossetia, as it gradually integrates the region militarily, politically, and economically with Russia. To secure the territory of South Ossetia, Russia routinely engages in a "borderization" strategy that Georgia calls a "creep-ing occupation." Since South Ossetia was not a formal administrative unit of Georgia, its boundaries were not clearly defined. Thus, Russian forces have taken it upon themselves to demarcate this largely unrecognized "interna-tional" border by erecting border fences and signs. As part of the strategy, on multiple occasions since 2013, Russia has literally shifted the boundary posts of the border to expand South Ossetia's territory.[13] For example, in a land

---

13. A report by the Heritage Foundation identifies ten cases of borderization from May 2013

FIGURE 1.2. Russia's borderization strategy in Georgia.

grab in July 2015, South Ossetia gained access to a mile-long stretch of the Baku-Supsa oil pipeline and pushed the border to within 500 meters of the E60 highway connecting Azerbaijan and the Black Sea (Figure 1.2). Media reports indicate that residents have gone to bed thinking that their house was located in Georgia proper and have woken up finding themselves living in South Ossetian territory.[14] Slowly but surely, these Russian encroachments into Georgian territory continue unabated, despite objections from Georgia and the West.

As Russia pursues its borderization strategy in Georgia, it has also utilized unconventional methods to expand its territorial reach in another neighbor, Ukraine. Crimea is a strategically located peninsula in the northern Black Sea. Home to the Russian Black Sea Fleet and a largely Russian-speaking population, Crimea had been part of Russia from 1783 to 1954, until the Soviet Union transferred control of the peninsula to Ukraine. In late February 2014, in the midst of the Euromaidan protests in Ukraine, reports of sightings of "little green men" emerged in Crimea. These men in question were Russian-speaking armed soldiers wearing green military gear without any insignia. While it was commonly

---

to August 2017 that resulted in additional Georgian territory falling within South Ossetian territory (Coffey 2018).

14. North 2015.

understood that the men were Russian military personnel, likely including members of the 810th Marines Infantry Brigade and Russian special forces, the Kremlin denied the presence of any Russian military forces in Crimea. Instead, Russian President Vladimir Putin claimed that they were merely members of local "self-defense groups."[15]

These non-uniformed soldiers played a key role in Russia's annexation of Crimea. On February 27, armed men seized the parliament building in Crimea's capital, Simferopol, and raised the Russian flag.[16] In the following days, unmarked Russian soldiers occupied and blockaded the Simferopol airport and several military bases in Crimea.[17] With their help, Russia then moved arms and military equipment into Crimea and eventually took control of the peninsula. Under Russia's guidance, Crimean officials quickly scheduled a referendum on Crimea's political status, and within a month of the appearance of the "little green men," Russia annexed Crimea into its own territory. Following this success in Crimea, non-uniformed Russian troops appeared in eastern Ukraine to support the pro-Russian separatist forces in Donetsk. The use of such deniable forces appears to have become a standard tactic in Russia's strategic playbook.[18]

Like China, Russia has largely relied upon gray zone tactics to pursue its territorial ambitions in recent years. Russia's expansion into Georgia has largely consisted of a series of salami tactics. It has gradually consolidated its power in South Ossetia over the past quarter-century and is slowly expanding its reach into Georgian territory through its borderization strategy. Moving border fences to redraw borders though a "creeping occupation" is a strategy designed to incorporate small steps that will likely not trigger a significant reaction on their own. Russia's annexation of Crimea provides an example of a related tactic common in gray zone campaigns known as a fait accompli, or land grab.[19] In this case, Russia seized control of the entire territory, hoping that it would not provoke an armed response from other states. Through this combination of salami tactics and faits accompli, Russia has been able to strategically advance its interests without resorting to open warfare.

---

15. Shevchenko 2014.

16. Booth and Englund 2014.

17. Oliphant 2014.

18. Altman 2018.

19. Altman 2017; Mazarr 2015; Tarar 2016.

Two Puzzles

In many ways, the expansionist behavior of China and Russia cuts against the general trend in international relations in recent decades. For one, since the end of World War II, the international community has largely embraced a norm against the use of coercion to revise territorial borders.[20] Historically, major powers routinely used their military might to try to expand their territorial control, and borders often shifted as a consequence of these conflicts. However, such coercive territorial revision is no longer widely seen as acceptable behavior. Additionally, in an increasingly interdependent global environment, uncertainty about borders can have greater economic consequences. Mutually recognized boundaries clearly define property rights and jurisdictional control, facilitating economic activity and international trade.[21] Thus, one might expect that these economic incentives would push countries to settle any outstanding disputes over territorial and maritime control. However, as China and Russia have become more integrated into the global economy in the first two decades of the twenty-first century, they have actively perpetuated these property rights disputes with the aim of expanding their territorial reach.

The expansionist activities of China and Russia are also notable in that they have, by and large, not escalated to major armed conflicts. The desire for territory has historically been one of the predominant causes of war between countries. However, both China and Russia have preferred to pursue tactics that remain in the gray zone between traditional diplomacy and war. Only during its brief five-day war with Georgia in 2008 has Russia engaged in significant active combat with a neighboring country, and China has largely avoided violent international military clashes to advance its claims in the South and East China Seas.[22] At the same time, both countries have deferred negotiating settlements over these territorial claims. By strategically delaying the resolution of boundary disputes, China and Russia have been able to pursue salami tactics to achieve their expansionist goals over time.

These patterns highlight two key puzzles. First, why do China and Russia pursue policies of territorial expansion in the face of widely accepted norms of territorial integrity and their increased interconnectedness with the global economy? Second, why have China and Russia largely relied upon gray zone tactics

---

20. Atzili 2011; Zacher 2001.

21. Simmons 2005.

22. The two notable exceptions to this were the brief naval disputes between China and Vietnam over the Paracel Islands in 1974 and the Spratly Islands in 1988.

to pursue these goals, rather than overt military force or negotiations through international institutions? For example, why has Russia not launched a full-scale war to occupy its neighbors, as Iraq did against Kuwait in 1990? On the other hand, why has China refused to settle its maritime boundary disputes in the East and South China Seas in an international court? The goal of this book is to solve these puzzles.

Market Power, Economic Interdependence, and International Institutions

To explain the behavior of states like China and Russia, we develop a theory of market power politics. This theory helps us to solve these puzzles by identifying factors that motivate and constrain the expansionist behavior of states involved in property rights disputes. Competition for market power can motivate states to take aggressive actions to expand their territorial reach or to prevent others from doing so. At the same time, economic interdependence and international institutions can place constraints on the strategies that states take to achieve these goals. To help set up our argument, let us briefly introduce these three key pieces of the puzzle: market power, economic interdependence, and international institutions. Then we can show how they fit together to form our model of market power politics.

Market Power Motivation

Market power is the ability to generate prices that diverge from what would result from a fully clearing competitive market. In normal economic exchange environments, prices are set by supply and demand conditions in the market. Thus, for all practical purposes, buyers and sellers are price takers. However, in markets of imperfect competition—like monopolies, monopsonies, and oligopolies—opportunities emerge for firms to gain price-setting abilities. For example, when a producer controls a significant share of a market, it may be able to use its market power to raise prices above the market price. On the other hand, if a buyer sufficiently dominates a market, it may be able to benefit from below-market prices. When firms set prices in their favor, they are able to generate extra profits, which are known as *rents*. Given the value of these potential rents, firms have strong incentives to accumulate market power.

While the ability to set prices is clearly valuable to firms, our analysis primarily focuses on the benefits that this market power can provide to leaders of states. Economically, the rents that firms generate from price setting can provide a source of revenue for the state. Additionally, market power in key markets can promote domestic political stability and provide leaders with international bargaining leverage. These economic and political benefits will likely be greatest when states own or have significant control over firms that have the ability to set prices for a key good in the state's economy. In these cases, states have strong incentives to take steps to increase their firms' market power. At the same time, other states will be motivated to prevent these states from capitalizing on market power opportunities. As we will see, this creates a competitive international environment that can potentially lead to armed conflict.

We mainly focus our attention on market power competition in hard commodity markets. *Hard commodities* are natural resources that are mined or extracted, such as energy resources and metals. In Part II of the book, we will take a close look at the market power motivations of three hard commodity producers: Iraq (oil), Russia (natural gas), and China (rare earth elements). Since the production of hard commodities typically requires a state to have access to resource reserves, market power in these markets is often a function of who controls the territory where these reserves are found and the supply routes by which the commodity is transported. Thus, if a state wants to increase its firms' market power or prevent another state from doing so, it may need to expand its territorial control. If the benefits of market power are sufficiently high, the state may be motivated to take aggressive steps to do so.

Economic and Institutional Constraints

While competition over market power provides a motivation for territorial expansion, states can be constrained from taking escalatory actions in pursuit of these goals. One such constraint is *economic interdependence*. Observers of international relations often equate economic interdependence with high levels of economic interaction between states. However, interaction does not necessarily imply interdependence. Instead, economic interdependence exists when the costs of exiting an economic relationship are high. These exit costs depend upon the availability of substitutes and the costs of adaptation. For example, if a state that relies upon natural gas for home heating has only one supplier of natural gas and cannot easily access alternative sources, exiting the relationship with that supplier would be very costly. Such economic interdependence between states increases the costs and risks of using violence. If armed conflict would result in economic exit, states that are dependent on one another will face potentially destabilizing economic costs that may be sufficient to deter them from escalating property rights disputes.

International institutions provide another constraint on aggressive state behavior. *Institutions* are sets of rules that govern human behavior.[23] To help achieve efficient outcomes in the global political economy, states have created a variety of international institutions that promote coordination and cooperation through the reduction of transaction costs. In the context of territorial conflict, institutions reduce the need for states to turn to violence by providing an avenue for states to peacefully resolve their property rights disputes. They do this in two primary ways. First, institutions provide rules to guide states on how they should allocate property rights. For example, international law articulates principles to follow when drawing territorial borders, while UNCLOS outlines a set of rules for determining the boundaries of maritime areas. Second, institutions provide procedures to help states reach durable settlements that are in line with those rules. When states are not able to reach a negotiated settlement on their own, they have the option of turning to an international organization to mediate the dispute or an international court to lay down a legal ruling.

In some situations, however, the presence of institutions can discourage dispute resolution. For one, institutional rules may limit the ability of states to achieve their political goals. For example, in the South China Sea dispute, a settlement consistent with the rules laid out by UNCLOS would likely require China to give up its claim to much of the sea. This politically unacceptable outcome makes China less willing to pursue an institutionally based settlement. Additionally, the difficulty of altering legally established boundaries raises the stakes of any settlement reached in a property rights dispute. This can reduce the willingness of states to make concessions over territory or submit their claims to an international court. Thus, some of the same factors that make institutions effective dispute-resolution mechanisms can make states less willing to reach a settlement in the first place. For this reason, the ability of institutions to constrain aggressive behavior in the face of market power opportunities will vary depending upon the strategic situation.

A Model of Market Power Politics

Figure 1.3 shows how these pieces fit together in our model of market power politics. When a state faces an opportunity to increase or preserve its firms' market power through the acquisition of territorial or maritime resources, it has incentives to take aggressive steps to do so. This expansionist motivation can create or exacerbate a property rights dispute between two or more states. The decision to

---

23. North 1990; Ostrom 1990.

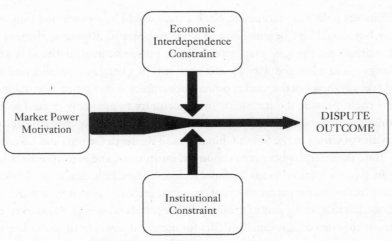

FIGURE I.3. Model of market power politics.

escalate these disputes is conditioned, however, by economic and institutional constraints. The economic constraints emerge from any interdependence that exists between the economies of the disputing states. These constraints can provide friction to slow down the dispute and limit one or more states' abilities to use force to obtain the market power opportunity. International institutional constraints work a bit differently. They can sometimes provide a similar friction if the structure of the institutional membership allows. However, their primary role is to provide more efficient, nonviolent pathways for states to resolve the dispute.

The outcome of the property rights dispute depends upon the combination of motivations and constraints that the states face. When economic and institutional constraints are low and the benefits of market power opportunities are high, states will be more willing to turn to violence. This can lead the dispute to escalate to war. Iraq's invasion of Kuwait in 1990 to gain the ability to affect oil prices provides a clear example of this path. At the other extreme, when disputants face significant constraints, they will be more willing to turn to peaceful dispute settlement. If there is a high level of economic interdependence and strong institutions can facilitate a mutually beneficial agreement, the economic benefits of resolving a property rights dispute may outweigh the potential gains from aggressively pursuing a market power opportunity. In these cases, we expect that disputants will likely agree to negotiated settlements or pursue legal dispute resolution through an international courts or arbitration panel.

In between these extremes, states with expansionist motivations may face a mix of strong and weak constraints. For example, a state may be economically dependent upon its rival but not find any of the available institutional

settlements politically acceptable. Such a state would be constrained from violence but would also be unwilling to agree to peaceful dispute settlement. In
these situations, the state may instead opt to pursue strategic delay. This delay
strategy could allow the state to wait until it is in a better bargaining position
and take advantage of the market power opportunity at that point. Alternatively,
delay could provide the state with an opportunity to gradually expand its territorial reach through gray zone tactics. Such strategic delay characterizes the
activities of China in the South China Sea and Russia in Georgia and Ukraine.

Thus, the interplay between economics, institutions, and expansionist behavior in international relations is multifaceted. The existing academic literature
largely focuses on the pacific effects of economic exchange and international institutions, but this is only part of the story. While interactions in global markets can
increase the costs of fighting, the desire for increased power in these markets can
motivate countries to expand their territorial reach. Similarly, strong institutions
can help countries effectively resolve disputes over property rights, but they also
limit the range of potential settlements and raise the stakes of any agreement that
is reached. Only by taking these competing incentives into account can we understand the varying strategies countries take to pursue their economic ambitions.

This helps us to solve the puzzles outlined in the preceding sections. Increased
economic interconnectedness and institutional developments have indeed
helped to reduce the propensity of the large-scale territorial wars of previous
centuries, but motivations for territorial expansion have not gone away. Among
these motivations is the desire for increased power in global and regional markets,
which can sometimes outweigh the economic benefits of settling disputes over
international boundaries. In some cases, countries are able to resolve property
rights disputes that arise from market power motivations through international
institutions. However, when major powers in this economically interconnected
world are constrained from using violence and cannot reach their market power
goals through an institutionally prescribed settlement, they instead turn to gray
zone tactics. Pursuing a strategy of delay, these countries use these tactics to try to
shift the strategic environment in their favor and gradually achieve their expansionist goals over time.

Contributions of the Book

In this project, we set out to understand some emerging patterns in the conflict
behavior of key players on the global stage, and we find that an important piece
of the puzzle lies in disentangling the economic and institutional incentives that
these countries face. For this reason, the discussion in this book cuts across many
of the established research communities within the field of international relations,

from conflict and security studies to international organization to international political economy. The contemporary academic environment encourages and values specialization, and this has undoubtedly led to a great accumulation of knowledge that has pushed the study of international relations forward. As a consequence, though, these separate research communities have fewer and fewer opportunities to speak to each other. We believe that this can hinder our ability to tackle many of the pressing issues that we face in international relations today, which often requires us to take a broader perspective. For that reason, our research necessarily draws from and speaks to a wide range of literatures in the field. We briefly highlight three examples of that dialogue here, leaving a more extended discussion of the contributions and implications of this research to the book's conclusion.

First, we provide new insights into the relationship between economics and international conflict. Scholarship in this area has largely focused on whether trade and economic interdependence reduce the likelihood of war. Following in the footsteps of Immanuel Kant, proponents of the liberal peace argue that trade dampens the prospect of international conflict.[24] Others challenge this claim, arguing that the relationship between trade and war is logically indeterminate or at least more complicated than a simple negative relationship.[25] Here we show that a different economic incentive—the desire for market power—can be an instigator of conflict. However, we also build upon the insights of the literature on the liberal peace by integrating the constraining effects of economic interdependence into our theory to identify contexts where competition motivates conflict but interdependence constrains one or more parties from escalation to war. By exploring the market factors underlying strategies of territorial expansion, we shed light on the critical nexus between territory, trade, and war.[26]

Second, we provide a unique analysis of the role of institutions in the context of territorial disputes that coincide with market power opportunities. In so doing, we reveal the pernicious incongruence between promoting efficiency and avoiding market power ambitions. The existing literature tends to focus on the ability of international institutions to help states efficiently resolve disputes.[27] However, we show that the development of such institutions can have

---

24. Polachek 1980; Russett and Oneal 2001.

25. Barbieri 1996; Chatagnier and Kavaklı 2015; Crescenzi 2005; Gowa 1995; Morrow 1999.

26. Schultz 2015.

27. In the context of property rights disputes, examples include Allee and Huth 2006; Gent and Shannon 2010; Huth, Croco, and Appel 2011; Mitchell and Hensel 2007; Powell and Wiegand 2014; Simmons 2002.

unintended consequences. In particular, institutions can constrain disputants to such an extent that they may be unable or unwilling to reach a dispute settlement. Many international institutions developed in the twentieth century were designed to add efficiency and transparency to international interactions. This focus on efficiency often comes at a price, however, if these institutions are unable to accommodate attempts to reshape markets. As a result, these institutions provide suboptimal solutions for disputing states when market power opportunities exist.

Third, our analysis of strategic delay in economically motivated territorial disputes provides a new understanding of the factors that lead countries to pursue gray zone strategies that lie between traditional diplomacy and war. Both experts in and practitioners of US national security policy have cited the emergence of gray zone campaigns as a critical concern in the contemporary global environment.[28] However, while the "gray zone" may have become a buzz word in Washington, the academic community has somewhat lagged behind in investigating these phenomena. Recently, international relations scholars have begun to put more theoretical rigor into our understanding of gray zone strategies.[29] Our research aims to contribute to the emerging literature on gray zone tactics by examining the economic motivations that can underlie the decision to turn to these tactics. Additionally, we show how the structure of international norms and institutions may be unexpectedly encouraging the pursuit of gray zone campaigns. Thus, our research on the strategic delay tactics pursued by Russia and China will be of interest to academics and policymakers alike.

Overview of the Book

The remainder of the book is primarily laid out in two parts. In Part I, consisting of Chapters 2 and 3, we lay out our theory of market power politics. Here we explain how market power motivations, in combination with economic and institutional constraints, influence countries' strategies in territorial disputes. In Part II of the book, we explore how these factors play out in the empirical world. In these chapters, we trace how market power motivations led to armed conflict between Iraq and Kuwait, expansionist gray zone tactics by Russia in neighboring Georgia and Ukraine, and strategic delay in the maritime disputes in the East and

---

28. Brands 2016; Green et al. 2017; Mazarr 2015.

29. Altman 2017; Altman 2018; Cormac and Aldrich 2018; Green et al. 2017; Lanoszka 2016; Mazarr 2015; Tarar 2016; Zhang 2019.

South China Seas. We then conclude with a discussion of the implications of our research for both scholars and policymakers.

Chapter 2 sets the stage for our theory of market power politics. Before we can analyze how the desire for market power can drive states to seek big changes in international relations, we must first map out the way things work in what we call *normal economic exchange* environments. In such environments, economic exchange takes place in perfectly competitive markets where both buyers and sellers are price takers. Whenever this characteristic holds, states have an interest in using institutions to maximize coordination and efficiency. These international institutions facilitate global commerce by reducing the transaction costs that political and economic actors face.

Territorial and maritime boundaries promote economic production and trade by establishing and clarifying property rights between states. International property rights disputes arise when states disagree over the sovereign control of territorial and maritime resources. Resolving these disputes can provide significant mutual economic benefits to the states involved, but it is often difficult for states to do so on their own. In normal economic exchange environments, international institutions can help resolve these territorial and maritime disputes by providing rules for allocating property rights and procedures to reach durable settlements. We illustrate the process of institutional management with a brief discussion of the resolution of the maritime boundary dispute between Romania and Ukraine.

In Chapter 3, we establish an argument for why property rights disputes that contain market power opportunities are particularly difficult for institutions and states to navigate, leading to interesting and important outcomes such as strategic delay and even the onset of war. To understand the emergence of war or delay in these disputes, we need to understand why states would be motivated to engage in non-cooperative behavior. One such motive is economic competition. When states are faced with an opportunity to achieve a significant increase in market power through the acquisition of territorial or maritime resources, they may have incentives to use force to seize the territory in question. This can lead these disputes to escalate to military conflict.

However, while states may have motives to achieve market power through violence, they are often constrained from doing so. We identify two forms of international constraints that shape decisions to pursue market power opportunities: economic interdependence and international institutions. When one state is economically dependent upon another, it is unwilling to incur the economic costs that would result from an escalation of conflict. Additionally, as we discuss in Chapter 2, institutions can help states effectively resolve disputes. However, when institutional rules and procedures prevent states from achieving market power opportunities, the same factors that make these institutions so effective can sometimes deter states from pursuing institutional solutions in the first place. When considered in tandem, the set of constraints created by economic inter-dependence and incompatible international institutions can help us understand when and why states seem to wait in a holding pattern we describe as strategic delay. Such a strategy allows states to bide their time until they are able to resolve property disputes in their favor or engage in gray zone tactics that allow them to gradually achieve their market power goals.

Chapter 4 begins Part II and transitions the book from theory to empirical evaluation. Here we explain and justify our empirical strategy and case selection. First, we explain the motivation behind our use of qualitative cases as plausibility probes. Next, we outline the criteria that we use to select our cases. In particular, we focus our attention on hard commodity markets and limit our analysis to states that have significant control over their firms in these markets. We then introduce our three empirical foci: Iraq's invasion of Kuwait in 1990 and its desire to become a price setter in the global oil export market; Russia's territorial expansionism and contestation in Georgia and Ukraine as a result of its desire to preserve Russian market power in the regional natural gas market; and China's extended territorial expansionism in the South and East China Seas in the context of competition over seabed resources, including rare earth elements.

In Chapter 5, we examine how a desire for market power led to the decision of Iraq to invade Kuwait in August 1990. This case highlights the use of force to improve Iraq's position within the global oil market at a time when it desperately needed new resources. The existing institutions governing oil export revenue, such as the Organization of the Petroleum Exporting Countries (OPEC), could not provide effective solutions for Iraq's needs following a decade of war with Iran. Given the absence of normal political solutions, Saddam Hussein explored the annexation of Kuwait and its oil reserves. This opportunity to improve Iraq's oil position worked in tandem with the classic commitment problems that plague international bargaining environments, leading to conflict. Kuwait was able to fend off the attack only with the help of the world's major oil consumers, who stood to lose from Iraq's increased market power. The United States and its part-ners feared that Saddam Hussein's pledge to not invade Saudi Arabia was not credible, and the subsequent shift in market power for Iraq in the event of seizing Saudi oil reserves would represent a major transition for Iraq to a price setter for global oil.

The Coalition of the Gulf War formed and successfully expelled Iraq from Kuwait's borders. But even after Hussein learned that the United States and Operation Desert Storm would successfully compel his forces back home, he maintained his strategy of using force to revise Iraq's competitiveness in the oil market. In the wake of the initial conflict, when Iraq made a hasty withdrawal,

Hussein's army endeavored to irrevocably alter the competitiveness of Kuwait by dismantling the Kuwaiti oil infrastructure. This destruction was economic warfare by Iraq. While Hussein's first preference was to horizontally integrate this infrastructure and resources into the Iraqi economy, his second preference was to destroy it. Compared to leaving Kuwait intact, this option had the potential to improve Iraq's power in the oil market. The sanctions imposed by the international community in the wake of the war likely undid any of this advantage, but Hussein could not have predicted these costs accurately. As such, this case represents the causal path from market power opportunities to war.

In Chapter 6, we explore how market power incentives in the regional natural gas market have shaped Russia's foreign policy activities since 2000. Russia is the largest exporter of natural gas in the world, and the Russian economy is largely dependent upon the production of hydrocarbons. Most of Russia's gas exports flow to Europe, and Russia would greatly benefit if it could monopolize Europe's consumption of natural gas and thus control the regional market price for gas. This would give Russia the ability to extract rents and use its price-setting capability as a lever for political gains. However, Russia's position in the European gas market is threatened by a number of factors, including the politics of transit states, the emergence of competitive suppliers, and changes to the pricing structure of European gas.

In light of these challenges, Russia has pursued a multipronged strategy to maintain its market power. Part of this strategy involves moves by the Russian gas company, Gazprom, to increase its control of the supply chain through vertical integration and the construction of bypass pipelines. However, most importantly for our discussion here, Russia has also pursued a policy of territorial expansion in Georgia and Ukraine to achieve its market power goals. By increasing its presence in the breakaway regions of South Ossetia and Abkhazia and engaging in a gradual "creeping occupation" of territory in Georgia, Russia has moved to destabilize a critical transit state for gas exports from Azerbaijan. On the other hand, Russia's annexation of Crimea and intervention in eastern Ukraine have opened up additional offshore gas reserves in the Black Sea and have disrupted Ukraine's ability to increase its own domestic gas production. In these disputes, Russia has largely pursued strategic delay, which has allowed it to gradually expand its territorial reach through salami tactics. Economic interdependence constrains Russia from pursuing significant military escalation, and international institutions do not provide a viable avenue for Russia to achieve such territorial gains.

In Chapter 7, we turn our attention to the ongoing property rights disputes in the South and East China Seas. Over the last half-century, China has increasingly invested in its ambitious territorial and maritime claims in both the East and South China Seas. By making and pressing these claims, China has entered into prolonged disputes with Japan, the Philippines, Vietnam, and other states. An ability to gain exclusive rights to mine the deep seabed in the East and South China Seas would preserve China's current power in the rare earth elements market, ensuring that China will have steady and affordable access to these minerals for its downstream economies focusing on high-end electronics manufacturing and green-energy infrastructure. Preserving this market power is key to reducing risk as China pursues its "Made in China 2025" economic policy. With the International Seabed Authority's recent development of a mining code that prioritizes the environmental concerns surrounding deep seabed mining for rare earths, China is reluctant to cede authority to this international institution.

Yet, full throttle territorial expansion has been slow to arrive in the South China Sea, and China has shown little indication that it seeks to force the issue with Japan in the East China Sea. Asymmetric economic interdependence between China and its South China Sea rivals prevents the Philippines and Vietnam from escalating their territorial disputes militarily. Without fear of escalation, China has pursued a practice of strategic delay in which it resists conflict resolution by international regimes and slowly but determinedly moves to consolidate de facto control of the South China Sea through salami tactics. Similarly, in the East China Sea, both China and Japan have pursued a practice of strategic delay. As China has become emboldened in the last decade, Japan has used the combination of economic and institutional constraints to prevent the resolution of this dispute in China's favor, which could lead to a spike in Chinese market power. Japan's key role as a trading partner and contributor of foreign direct investment prevents China from escalating to more aggressive tactics, while the design of institutions such as UNCLOS makes legal dispute resolution too risky for either side.

In the concluding chapter, we return to our theoretical road map and the broad strokes of our empirical analysis to examine the contributions of the book and the lessons it offers for scholars, students, and practitioners of world politics. To complement our extended case studies of violence and strategic delay, we provide a brief discussion of Russia's decision to abandon strategic delay and agree to a settlement of the long-running dispute over the Caspian Sea. We also outline a set of questions for future research on market power politics. Then, to reflect upon how our research informs our understanding of international relations, we contemplate the effects of market structure on conflict behavior, the limitations of international institutions, and the future role of gray zone tactics by countries like Russia and China. We conclude the book with a discussion of some of the policy implications that follow our study of market power politics.

PART I

A Theory of Market Power Politics

2

Markets, Institutions, and Property Rights Disputes

OUR MAIN GOAL in this book is to understand when and how economic competition drives states to seek big changes in international relations. Before turning our attention to these rare but consequential situations, we want to first consider how politics normally works in the global economy. This will allow us to set the stage and provide a general overview of how markets and institutions generally shape the behavior of global actors like states and firms. The typical state of affairs in international relations provides a baseline from which we will be able to evaluate the market power opportunities and expansionist behavior we explore later in the book.

Of course, even a casual observer of global events may question the idea that we can define what constitutes "normal" politics in international relations. The world around us is constantly evolving, sometimes in unforeseen ways. Due to factors like globalization and technological changes, many aspects of our lives today look very different than they did even a few years ago. Moreover, media reports highlight singular and unexpected events—terrorist attacks, trade wars, Brexit, the coronavirus pandemic—that seem to defy any sense of normality. It can often seem as if international relations are characterized by a series of unprecedented "breaking news" events.

However, despite the headlines, most day-to-day activities in the global political economy are relatively mundane and follow relatively predictable patterns. Firms and individuals consistently trade billions of dollars in goods and services across borders every day. Global prices largely reflect supply and demand in relevant markets. States generally follow international law and abide by their treaty obligations. Our goal here is to provide a theoretical framework—largely based upon new institutional economics—that helps us describe and understand these general patterns. We recognize that the world does not always behave exactly in line with the theoretical ideal we outline here. However, it provides a good first approximation of the global political economy and a useful starting point for developing a theory of market power politics.

In this chapter, we characterize three general aspects of the international political economy. First, we describe the perfectly competitive markets that constitute what we call *normal economic exchange (NEE) environments*. In these environments, prices are set by supply and demand in the market. Thus, we can view both buyers and sellers in the NEE as being *price takers*. Since global actors cannot control the prices themselves, political behavior in the NEE largely aims to coordinate or manipulate the effects of the market price. In this way, the market largely sets the agenda for politics, and efforts to insulate actors from global markets are often short-lived.

Second, we consider the role of institutions in fostering cooperation and promoting trade in NEE environments. International institutions are rules that constrain the behavior of states, firms, and other global actors. These actors create institutions to help achieve efficient outcomes, largely through the reduction of transaction costs. Over the past century, institutions like the General Agreement on Tariffs and Trade (GATT) and the World Trade Organization (WTO) have evolved to facilitate economic cooperation and growth in global markets. Institutions also provide a mechanism for promoting market efficiency through the creation and maintenance of property rights. At the global level, territorial borders and maritime boundaries provide an institutional means for establishing and clarifying international property rights among states.

Finally, we discuss the potential for institutions to help states resolve territorial and maritime disputes that may threaten the efficiency of NEE. We call such disagreements between states over the sovereign control of territorial and maritime resources *(international) property rights disputes*. At their heart, property rights disputes involve competing claims over the location of international boundaries, which can create economic inefficiencies. Thus, resolving these disputes should provide significant mutual economic benefits to the states involved, but states sometimes face barriers to reaching a successful agreement. International institutions can assist states in overcoming these obstacles by providing rules for drawing boundaries to allocate property rights and procedures for reaching durable settlements. We illustrate this process with a description of the resolution of a maritime boundary dispute between Romania and Ukraine.

Normal Economic Exchange: A Market of "Price Takers"

The vast majority of events in world politics occur in NEE environments. Indeed, these events are so mundane that we do not often think about them. We sometimes aggregate them into yearly statistics or group them into broad categories. For example, the exchange of goods and services between firms and individuals in the United States and Canada is constant. Investments, products, and services traverse the border every day, so much so that we pay little attention. More broadly, states in the international system exchange goods, capital, and services at a tremendous rate. We often evaluate these flows in aggregate form, examining patterns of coordination and barriers to cooperation and searching for ways to improve the efficiency of global transactions. Or we focus on the political impact of this economic exchange, highlighting the winners and losers of globalization to show how the global economy affects our political lives every day. Having a sense of how things work in this NEE environment will be helpful as we explore how a desire for market power can sometimes derail normal politics. At first glance, describing what a political economy scholar would deem to be a normal economic story seems like a daunting task. Entire courses can be taught about international economics, tracing the evolution of exchange from barter to prices, from mercantilism to Ricardian comparative advantage. One would then examine modern international trade theory and general equilibrium models, comparing Heckscher-Ohlin to Ricardo-Viner approaches to see if the interesting politics falls along divisions based on factors or sectors. We assume the reader has some familiarity with these basics, but it is not strictly necessary for digesting the arguments in this book. All we wish to do here is emphasize one key characteristic of what is typically labeled the standard trade model: when economic exchange takes place in perfectly competitive markets, both buyers and sellers are *price takers*. Whenever this characteristic holds, states have an interest in using basic institutions to maximize coordination and efficiency. The extreme version of this condition is the competitive paradigm that defines late-twentieth-century neoclassical economics.

This competitive equilibrium, or Walrasian equilibrium, is characterized by the existence of many players and flexible markets. The existence of such an equilibrium requires two definitional requirements.[1] First, demand must equal supply, meaning that markets clear. Second, consumers maximize their utility subject

---

[1]. Ellickson 1993, 15. The label honors Leon Walras, who contributed to the development of the general equilibrium theory.

to a budget constraint. When translated to international economic exchange and the notion of trade, economists still rely on the platform of a competitive equilibrium when modeling standard trade theory. This is, of course, an assumption that is rarely fully observed in the real world, but what matters is the extent to which it holds. In competitive, Walrasian equilibria, buyers and sellers are plentiful. A firm selling a good has many buyers to choose from, and buyers have many sellers seeking their business. Entrance into (and exit from) the market is fairly easy, ensuring a steady stream of agents on both sides of the transaction.

In such an environment, prices are set based on supply and demand conditions within the market. Goods and services are bought and sold at the price that the market will bear, and as such we can say that the market sets the price. As prices are set by the market, buyers and sellers react to this information accordingly. In such situations, we refer to both the buyers and sellers as price takers. This is not to say that everyone is satisfied with the market price, nor do we intend to make some sort of preferential statement about markets. We simply focus on the notion that when markets approach perfect competition and neither buyers nor sellers can control the price of goods and services, politics relating to economic exchange centers around this equilibrium.

Political behavior in the NEE typically involves attempts to coordinate or manipulate the effects of the market price, since the price itself cannot be manipulated. Sellers and manufacturers, for example, may lobby their government to raise trade barriers in an attempt to compensate for disadvantages resulting from being uncompetitive on the open market. In the United States, for example, we have historically observed efforts by the automobile, steel, and energy industries to lobby the federal government to restrict imported cars, steel, and energy. These industries (and their governments) typically cannot affect the global market price for cars, steel, or energy, but they may be able to use the apparatus of the state to insulate themselves from market pressures.

Such political efforts have two things in common. First, the market defines prices and sets the agenda for politics. Politics, therefore, can be thought of as a reaction to the market price in an attempt to help or gain support from domestic constituencies. Second, even successful lobbying efforts in the NEE tend to be temporary. Because these efforts create local incentives or insulation from the market, they are inefficient from a macro-economic perspective. That inefficiency creates a tension (e.g., between a market price for steel and the protected domestic price) that is politically difficult to sustain.

Institutions play a central role in shaping the political behavior of states and firms in the NEE. Many of the institutions in world politics are designed to facilitate coordination and cooperation in this mundane dimension of political economy. For example, the Bank of International Settlements (BIS) was formed in 1930 (and resurrected after World War II), and its main job is to serve as a coordinating institution for central banks. Originally designed to facilitate repayments after World War I, the BIS has evolved to become a central bank for central banks, focused on financial and monetary stability and facilitating interstate transactions. The BIS is an excellent example of the role of institutions in facilitating coordination across the global economy in part because we rarely think about it or hear about it in the news. It serves an important role in facilitating economic exchange, but it does so quietly. Of course, the BIS provides just one example of the many institutions created by states over the past century to regulate the international political economy. We now turn our attention to these institutions and the role that they play in global economic relations.

The Role of Institutions in International Politics

Institutions are sets of rules that govern human behavior.[2] These rules range from formal laws to informal conventions. By forbidding, requiring, or permitting particular types of behavior, institutional rules alter the incentive structures of actors. In this way, institutions constrain choices that are made by individuals, organizations, states, and firms. Thus, political and economic outcomes are, in part, a function of the institutional rules that actors face. Knowing this, strategic actors create and change institutions with the hope of reaching better outcomes than they otherwise would.

In international relations, states often design institutions aimed to achieve efficient outcomes.[3] When we say that one outcome is efficient, we mean that there is no other outcome that makes at least one actor better off and makes no one worse off.[4] Thus, the goal of these institutions is to avoid suboptimal outcomes (at least in the eyes of those who create the institution). The canonical example of an environment where institutions can promote efficient outcomes is a prisoner's dilemma situation in which an institution can help states achieve mutual cooperation when they would otherwise have a dominant strategy to defect. However, institutions can also benefit states in a variety of other strategic settings. For example, in a bargaining situation, institutions can help states overcome obstacles to reaching a mutually beneficial agreement. Institutions can also provide focal points that allow states to more quickly and easily coordinate their actions.

---

2. North 1990; Ostrom 1990.

3. Keohane 1984; Koremenos, Lipson, and Snidal 2001.

4. This is technically known as Pareto efficiency.

In the NEE, the purpose of many international institutions is to promote more efficient economic exchange by fostering cooperation and facilitating trade. The primary way that they do this is by reducing the transaction costs that political and economic actors face. Such transaction costs include search and information costs, bargaining and decision costs, and policing and enforcement costs.[5] These costs are generally the result of uncertainty, which creates inefficiencies in the exchange of goods and services and results in a reduction in overall economic welfare. International institutions can thus reduce transaction costs by providing information to allow actors to make efficient choices. In doing so, they can facilitate bargaining between states and between firms, and they can provide mechanisms to allow actors to monitor compliance with agreements and coordinate enforcement.

Most of these institutions are familiar to us, such as the WTO and its predecessor, the GATT. A quick look at the GATT illustrates the pursuit of coordination and efficiency. Participation in the GATT required signing on to the notion of Most Favored Nation status for the other participants, meaning trade barriers could be established at the good or industry level and then applied equally across all member states. Trade barriers were then reduced through an evolution of norms and contracts established through rounds of negotiation, with mixed success. Of course, equal implementation is a matter of politics and power, and we do not need to adhere to some naive notion that the GATT was good for all states and firms all the time. It is enough to observe that states and firms worked within the GATT framework, manipulating it to their advantage when they could, but abiding by the fundamental rules that made the GATT work.

When the WTO emerged in 1995 from the Uruguay Round negotiations as the successor to the GATT, it was an evolution of the institution designed to improve its ability to reduce uncertainty and improve transparency. Four of the six core principles that guide the WTO deal with efficiency and coordination. These include opening borders for trade, making trade barriers more transparent and predictable, making markets more competitive, and decreasing discriminatory trade practices.[6] In this way, the WTO is designed to make the normal economic environment work more efficiently, promoting competitive markets while reducing barriers to trade.

Perhaps the most interesting dimension of the WTO is its Dispute Settlement Body (DSB). The DSB provides a process whereby member states can bring trade

---

5. Dahlman 1979.

6. The last two principles are to improve the economies of less developed nations and to protect the environment.

disputes to the WTO for policy recommendations. For each dispute, a dispute panel reviews the case and makes a recommendation. Once the panel has made its recommendation and any appeals have taken place, the panel recommendation is considered accepted by the DSB unless a consensus of member states vote against acceptance. This structure makes the panel's recommendations more likely to stand. Importantly, the WTO has no authority to implement or enforce a panel's recommendations. The DSB simply approves or rejects the panel's policy decision, which for example authorizes the member state that brought the claim to impose sanctions against the state that is breaking the rules. The WTO brings no sanctions of its own, nor does it enforce the sanctions behavior of member states. Put another way, even in dispute settlement, the WTO serves in a coordination and informational capacity.

Institutions and Property Rights

The importance of property rights as a foundation for efficient markets is so well accepted in the competitive paradigm as to be largely taken for granted. Skaperdas, for example, views his investigation of contests within incomplete property rights systems as a departure from the neoclassical competitive equilibrium.[7] In the study of international political economy, political institutions are often presented as a mechanism to achieve the efficiencies of competitive markets through the creation and maintenance of property rights.[8]

Institutions can also facilitate cooperation in the NEE through the establishment of property rights. Efficient economic exchange requires a clear, accepted allocation of property rights. When property rights are unclear, the exchange of goods is riskier. A buyer may be unsure as to whether the seller has the authority to transfer ownership of a good to her. Thus, once an exchange is made, a buyer may be uncertain as to whether she will be able to maintain a property right over the good. Similarly, when firms are uncertain about their property rights, investment in economic production will be riskier. Thus, uncertainty about property rights can create significant transaction costs. Institutions can reduce these costs by clarifying the allocation of property rights among actors and protecting those rights.

---

7. Skaperdas 1992.

8. Keohane 1984; Simmons 2000b.

In international relations, territorial borders and maritime boundaries are institutions that help to establish and clarify property rights.[9] Borders are jurisdictional rules that divide territory and any associated economic resources among states. By creating a border, states can identify a clear physical area over which they maintain sovereignty. They are then able to clarify the property rights of firms and individuals within their territory. If these borders are unclear or disputed, firms face significant jurisdictional and policy uncertainty, which makes production and trade riskier. Clear, mutually recognized borders, on the other hand, provide these economic actors with information about which state has jurisdictional control in a given territory and which policies they need to follow, leading to a reduction in transaction costs. Thus, mutually recognized borders should facilitate and promote economic activity and international trade.[10]

Maritime boundaries delimit sovereign control of potentially valuable resources found within and below the world's oceans and seas. While the concept of recognized borders that demarcate the territory under sovereign control of a state has long been established in the modern state system, the contemporary institutional structure of maritime boundaries has more recent origins. The United Nations Convention on the Law of the Sea (UNCLOS) was signed in 1982 and went into effect in 1994.[11] Among its many provisions, the Convention outlines the areas to be defined by maritime boundaries, including territorial seas and exclusive economic zones (EEZs). A state's territorial sea is considered part of its sovereign territory. On the other hand, an EEZ is an area in international waters in which the state has a sovereign right to all economic resources below the surface of the sea. Mutually recognized boundaries of states' territorial seas and EEZs help to define property rights over economic resources in many of the same ways that territorial borders do. This should, in turn, reduce transaction costs and lead to an increase in production and trade.

In sum, there are a wide range of international institutions that structure the political economy in the NEE. These institutions play an important role in facilitating commerce on a global level. By reducing transaction costs, formal institutions such as the WTO decrease barriers to trade and ideally increase international economic exchange. Territorial and maritime boundaries can also

---

9. This conceptualization of borders as institutions was first proposed by Simmons 2005. The discussion here largely follows the framework that she laid out.

10. Simmons 2005.

11. As of 2016, UNCLOS has been ratified by 166 states and the European Union. The United States is not a party to the Convention, but it recognizes most of the principles of UNCLOS as part of customary international law.

be viewed as international institutions that promote economic production and trade by establishing and clarifying property rights between states. However, institutions not only facilitate economic cooperation in day-to-day interactions; they can also potentially help states manage crises that can threaten a shift in the entire market structure. We now consider institutional management of one such type of crisis: disputes over property rights.

Institutional Management of Property Rights Disputes

International property rights disputes are disagreements between states over the sovereign control of territorial and maritime resources. In territorial disputes, multiple countries lay claim to the same piece of land. Prominent examples of such disputed territory include Crimea, which is claimed by both Russia and Ukraine, and Kashmir, which is claimed by both India and Pakistan. Similarly, maritime disputes arise when states' claims over maritime areas—such as EEZs or continental shelves—overlap. To complicate matters, many maritime disputes are intertwined with territorial disputes over the control of islands, such as the Senkaku/Diaoyu Islands in the East China Sea and the Paracel and Spratly Islands in the South China Sea.

At their heart, these property rights disputes involve competing claims over the location of international boundaries. As we explained earlier, boundaries can be seen as international institutions that clarify property rights. When these boundaries are contested, economic inefficiencies emerge. Given the salience of the issues involved, disputes over territory are often contentious and can be a catalyst for costly armed conflict.[12] Moreover, these disputes can pose significant economic opportunity costs for states. Time and energy spent on addressing the dispute cannot be spent on more productive activities. Additionally, boundary disputes generate jurisdictional and policy uncertainty for economic actors, which puts a damper on trade flows. Unclear property rights also increase the risks of investment and infrastructure development for firms, inhibiting economic production. Thus, resolving property rights disputes should provide significant mutual economic benefits to the states involved.

To successfully resolve these disputes over property rights, states need to do two things. First, they must reach an agreement over any contested territorial or maritime boundaries. By mutually agreeing to these boundaries, they can clarify the allocation of corresponding property rights. Second, they must be willing to

---

12. Holsti 1991; Vasquez 1993.

commit to respecting these boundaries and property rights in future interactions. An agreement is not helpful if it is short-lived. By ensuring long-term compliance with the terms of the boundary settlement, states will hopefully be able to sustain a higher level of economic cooperation going forward.

It is often difficult for states to achieve these goals on their own, as they will often be reluctant to make the concessions that would be necessary to reach a lasting settlement. Domestic constituencies, especially those with economic or cultural ties to the contested areas, may pressure leaders to avoid backing down from territorial or maritime claims. Additionally, a leader who concedes control of resources to a rival may be perceived as weak, which could open up the possibility of future challenges. Finally, and most critically for our discussion in the next chapter, a settlement may require a state to give up its efforts to achieve a significant shift in market power.

International institutions provide a potential mechanism for states to peacefully resolve disputes over property rights. Earlier we discussed how some international institutions structure political and economic relations in the NEE. Here we discuss how institutions can help states resolve crises that threaten a shift in market structure. As before, the promise of international institutions is that they can provide gains in efficiency. In this case, institutions facilitate the resolution of property rights disputes by reducing the transaction costs of reaching an agreement on the distribution of territorial and maritime resources. In doing so, they promote the establishment of clear, mutually recognized boundaries that will foster more efficient economic exchange in the long run.

We should emphasize that the institutions we consider here are distinct from the boundaries themselves. Otherwise, our argument would run the risk of being circular in nature. In the previous section, we introduced the idea that mutually recognized territorial and maritime boundaries function as international institutions that reduce transaction costs. Our focus here is instead on those institutions that help states define the location of international boundaries. Thus, while the boundaries perform the first-order tasks of clarifying property rights and jurisdictional control over territory, these institutions perform the second-order task of helping states choose, implement, and maintain an effective boundary among the potential boundaries that could be drawn.[13]

There are two key features of international institutions that promote the settlement of property rights disputes. First, institutions provide rules to guide the allocation of property rights. For example, these rules may outline what an appropriate territorial or maritime boundary should look like. Second, they provide

---

13. Knight and Johnson 2011, 19.

procedures that states can use to resolve these disputes nonviolently. These procedures, which include adjudication and arbitration, provide a mechanism for states to reach settlements that are consistent with the rules laid out by the institution.[14] In following sections, we describe in more detail these processes by which institutions structure bargaining over property rights.

Rules for Allocating Property Rights

As we have noted, institutions are sets of rules that structure global political and economic behavior. To resolve property rights disputes, states need to coordinate on and then commit to a particular allocation of property rights. This typically requires disputants to agree on the location of territorial and maritime boundaries. International institutions can provide rules to help states identify how these boundaries should be drawn. International legal principles and provisions of UNCLOS are two examples of institutional rules that regulate the allocation of property rights in global politics.

A large body of international law articulates principles that should be used to resolve disputed claims over territory. For example, due to the principle of *pacta sunt servanda*, states are generally bound by previous border treaties that they signed and ratified. Additionally, the exercise of effective control of a territory can provide a state with a strong legal claim to a title over that territory.[15] In cases of colonial independence, the principle of *uti possidetis* presumes that the boundaries of a new state will follow the administrative boundaries of the previous colonial regime.[16] When states draw international borders along navigable rivers, the thalweg principle indicates that the boundary should follow the center of the main navigable channel of the river.[17] These are only a few examples of the many legal principles that provide rules for states to follow when drawing borders to resolve territorial claims.

Relatedly, UNCLOS provides rules for resolving maritime claims. As we noted earlier, UNCLOS outlines the rights and responsibilities of states in various types of maritime areas. It also specifies how the boundaries for these

---

14. In a technical sense, what we describe as rules and procedures could both be referred to as institutional rules (Martin and Simmons 2013). We choose to use the separate terms for the sake of clarity of exposition. Our use of the terms *rules* and *procedures* largely corresponds to what Duffield (2007) refers to as regulatory rules and procedural rules, respectively.

15. Grant and Barker 2009, 452.

16. Ibid., 655–656.

17. Ibid., 602–603.

maritime areas should be drawn. For example, a state's sovereign territorial sea extends 12 nautical miles from its coastal baseline. EEZs—maritime areas where states have exclusive rights over the exploitation of natural resources—extend beyond the territorial sea up to a distance of 200 nautical miles from the baseline. UNCLOS outlines the specific methods that should be used to identify the location of the baseline for measuring these maritime areas. It also articulates how boundaries should be drawn when two states' maritime areas would overlap. For example, in most cases, such territorial sea boundaries should follow the median line equidistant from each state's baseline. The rules on the delimitation of EEZs are less specific, but they require that these boundaries should be drawn "on the basis of international law . . . in order to achieve an equitable solution."[18]

Thus, international institutions provide sets of rules that states can follow when creating and clarifying territorial and maritime boundaries. In the anarchic international system, there is no police force to enforce these institutional rules. By and large, rather than coercing the activities of sovereign states, these institutions articulate the expected behavior of international actors. Given this, one might wonder why states would actually obey the directives of international institutions, especially when bargaining over highly salient issues such as territory. This is perhaps one of the central questions in the study of international institutions, and we do not aim to provide a definitive answer here. Instead, we highlight two critical mechanisms that may lead states to opt for institutionally prescribed settlements in property rights disputes: focal points and reputation costs.

Focal Points

International institutions can provide focal points for resolving disputes over property rights. Focal points allow actors to coordinate when multiple potential outcomes are available.[19] Institutional rules can identify particular outcomes or actions that international actors should choose. By providing this information, they can reduce the transaction costs of coordinating on an outcome. One often thinks of focal points as being useful in cases of pure coordination, in which all potential coordinated outcomes are similarly beneficial. For example, international agreements such as the Warsaw Convention harmonize standards for air

18. United Nations 1982, Article 74.1.

19. Schelling 1960.

travel and air safety across countries.[20] By coordinating on the same rules and regulations, such as the requirements for passenger tickets or baggage tags, countries can make international air travel more efficient. In such situations of relatively pure coordination, focal points are largely benign, as international rules do not create "winners" and "losers."

To be sure, a property rights dispute is not a situation of pure coordination. States have divergent preferences over what the allocation of property rights should be. All else being equal, each state would generally prefer to have control over more territory and economic resources. However, this does not mean that these are zero-sum situations. As noted earlier, disputing states have significant incentives to resolve uncertainty over property rights. Boundary disputes increase the potential for armed conflict and create significant opportunity costs for states. They also create uncertainty for economic actors, which can potentially lead to decreases in investment, production, and trade. Given this, resolving disputed territorial and maritime claims should generate economic benefits for the disputing states.

Thus, states attempting to resolve a property rights dispute have mixed motives. While competing states prefer to gain control of a greater share of the available economic resources for themselves, they also benefit from reaching any settlement that clarifies property rights and avoids the opportunity costs of an ongoing dispute. This is a prototypical bargaining problem. In such situations, focal points can still be useful mechanisms to help states coordinate on distributional outcomes. In the case of property rights disputes, if institutional rules indicate that a particular division of territory or a particular maritime boundary is most legitimate, disputants may see this outcome as a focal point upon which they should agree. Huth, Croco, and Appel argue that such focal points will emerge in territorial disputes when the relevant legal principles are unambiguous and clearly favor one side.[21] They find that when these conditions are met, disputants are significantly more likely to reach an agreement that settles a territorial claim.

Thus, the existence of institutional rules can potentially help states resolve property rights disputes by providing focal points. Of course, states will only be willing to agree and commit to a focal outcome when it makes them better off than a continued dispute. If this is true for all disputants, then the focal point would

---

20. The Warsaw Convention is formally known as the Convention for the Unification of Certain Rules Relating to International Carriage by Air. Guzman (2008, 26–27) highlights this as an example of pure coordination in international law.

21. Huth, Croco, and Appel 2011.

be self-enforcing. However, there are also situations in which an institutionally prescribed outcome may not be politically acceptable to one of the disputants. For example, a powerful state may be unwilling to make significant territorial concessions that are unpopular domestically or that would put it at a significant competitive disadvantage in the future. In such cases, disputants may eschew an institutional settlement of the dispute.[22] However, providing focal points is not the only mechanism by which institutions can influence state behavior.

Reputation Costs

In situations where a focal point is not self-enforcing, reputation costs can provide an additional means by which institutional rules can shape the resolution of property rights disputes. When a state violates institutional rules, it can gain a reputation for not adhering to its commitments. Such a reputation can prove politically costly to state leaders in their future interactions with international and domestic actors.[23] These reputation costs tend to be greater when institutional rules are formalized, as the institutional commitments of states are clearer and more visible and outside actors can more easily identify acts of noncompliance.[24] Thus, while there may be benefits to eschewing institutional rules in the short term, such actions can be potentially detrimental in the long term.

Reputation costs can emerge from various sources. At the international level, other states may punish the defecting state by being less likely to cooperate with it in the future.[25] International actors may be willing to impose such punishments in order to protect the credibility of the institution and the efficiency benefits that it provides. In turn, domestic audiences may also punish leaders who develop poor international reputations.[26] Constituency groups may withhold support for leaders who renege on international commitments that benefit them.[27] These domestic costs may be especially prevalent for governments based upon the rule of law, as domestic audiences may question whether a leader who flouts institutional

---

22. Gent and Shannon 2014.

23. Crescenzi 2018.

24. Abbott and Snidal 2000; Lipson 1991.

25. Guzman 2008.

26. McGillivray and Smith 2008.

27. Chaudoin 2014; Dai 2007.

commitments will be willing to respect domestic constitutional limits, including the protection of property rights.[28]

In the context of resolving property rights disputes, these reputation costs lead states to reach agreements that are in line with institutional rules. When states negotiate a solution to a boundary dispute, they face potential reputation costs if they actively push for a settlement that violates institutional rules. Additionally, they may face costs if they avoid pursuing an institutionally prescribed solution. Together these factors decrease the cost of agreeing to a settlement that corresponds to institutional rules relative to other potential settlements. Thus, reputation costs may open up opportunities for states to agree to institutional settlements that might otherwise not be achievable.

Once an agreement is made, states will also face greater reputation costs if they choose to renege on a settlement that is in line with institutional rules. Given this, we would expect that compliance with such settlements will be higher than other settlements. Thus, pursuing institutionally prescribed settlements can help states overcome potential commitment problems. This can make institutional settlements particularly attractive to a state looking for ways to sustain long-term economic cooperation. When we put this all together, the potential reputation costs that states face in the negotiation and compliance stages can help make institutionally prescribed solutions to property rights disputes possible.

Procedures for Dispute Settlement

In addition to laying out rules to guide the terms of agreements, institutions also provide procedures that states can use to resolve their disputes peacefully. In brief, these procedures provide a mechanism for states to reach settlements that are in line with institutional rules. While there are a wide range of institutional dispute resolution mechanisms in international relations, we focus here on two types of procedures that are used to resolve property rights disputes: adjudication and arbitration. In both of these procedures, disputants submit their claims to a third party who makes a legal ruling on how the dispute should be settled. The two forms differ in terms of the nature of the third party: adjudication is performed by a standing international court, while arbitration is typically carried out by an individual or ad hoc panel selected by the disputants.[29] In contemporary world politics, the most commonly used forum to adjudicate territorial disputes is the International Court of Justice (ICJ). Under UNCLOS, states can choose from

---

28. Simmons 2000a.

29. Cede 2009.

several institutional forums to resolve maritime boundary disputes, including the International Tribunal for the Law of the Sea, the ICJ, and ad hoc arbitration panels.

The use of these institutional forums can be a highly effective means to successfully resolve property rights disputes.[30] When states opt to pursue arbitration or adjudication, they do two things. First, they delegate decision control to a third party. Decision control refers to the "degree to which any one of the participants may unilaterally determine the outcome of the dispute."[31] Thus, the third party—rather than the disputants themselves—determines the terms of any dispute settlement. Notably, the decisions reached through arbitration and adjudication are generally based upon principles of international law. Second, the disputants pledge ahead of time to be legally bound by the decision reached by the third party. Thus, if a state chose to not abide by the ruling, it would also renege on this institutional commitment.

These factors help states reach settlements over disputed territorial and maritime boundaries. When states give up decision control to a third party, they are able to avoid the back-and-forth of the negotiation process. In bilateral negotiations, talks can fall through when one or both of the disputants walks away from the negotiating table before they reach an agreement. On the other hand, in arbitration and adjudication, a third party is able to lay down a settlement without any further consent from the disputants. In this way, these institutional forums can help states reach settlements of disputes over economic resources when they might not be able to do so on their own.

Of course, reaching an agreement is only the first step in resolving a dispute. The use of institutional forums can also lead to longer lasting settlements. The role of legality in institutional dispute-resolution processes can increase the costs of not complying with binding rulings. As discussed earlier, by reneging on a settlement based upon principles of international law, a leader may face reputation costs at the international or domestic levels. Such reputation costs may even be higher for violating settlements reached through an international tribunal since the state pledged ahead of time to be legally bound by the ruling. Reputation costs may be particularly relevant for adjudication by the ICJ or other dispute resolution bodies of international organizations. In these cases, member states may be more willing to enforce settlements produced by international organizations because failure to do so would raise doubts about the strength of the institution.[32]

---

30. Gent and Shannon 2010; Mitchell and Hensel 2007.

31. Thibaut and Walker 1978, 546.

32. Mitchell and Hensel 2007.

In addition to these increased international noncompliance costs, the use of institutional forums can also promote long-lasting settlements by providing domestic political cover for the leaders of the disputing states.[33] Suppose a leader is faced with the choice to comply with a settlement in which a state is required to make territorial concessions. A domestic constituency unhappy with the prospect of these concessions would have an incentive to punish the leader for complying with the settlement. If the concessions are the result of a bilateral agreement, the leader generally must take full responsibility for the terms of settlement. On the other hand, with arbitration and adjudication, the leader can shift the blame to the third party. Moreover, the domestic population may be more willing to accept concessions required by legal rulings, as they may be seen as being more legitimate. Therefore, due to this political cover, leaders may be more able to comply with settlements reached through institutional forums.

Despite the benefits of arbitration and adjudication, most property rights disputes are resolved outside of these institutional forums.[34] Legal dispute resolution can be a costly and lengthy enterprise. Given the uncertainty that it poses, states are often reluctant to give up decision control to an international court or arbitrator. Instead, states tend to prefer to try to settle their dispute bilaterally if they can, so they can have a greater say in the terms of any settlement. This is similar to the incentive for disputants in domestic legal settings to reach a plea bargain rather than deal with the cost of a court case or the risk of an unfavorable ruling.[35] Thus, empirically states generally only opt for arbitration or adjudication after they have failed to reach an agreement on their own.

Even in the case of bilateral negotiations, institutional procedures can still play a role in the settlement of property rights disputes. The ability to appeal to an institutional forum gives states with strong legal claims bargaining power in bilateral negotiations. Since these states can expect a favorable outcome from an institutional ruling, they can credibly threaten to turn to arbitration or adjudication if they do not receive sufficient concessions from a rival state. The rival state in turn may be willing to make such concessions in order to avoid ceding decision control to a tribunal. Thus, the expected outcome of arbitration or adjudication can shape the terms of settlements reached through bilateral bargaining.[36] This

---

33. Allee and Huth 2006; Simmons 2002.

34. Gent and Shannon 2011.

35. Landes 1971.

36. Manzini and Mariotti 2001.

provides an additional mechanism by which institutional procedures can guide states to allocate property rights in line with institutional rules.

In sum, institutions can help states resolve disputes by providing rules for allocating property rights and procedures to reach durable settlements. In the absence of strong institutions, peaceful crisis management can be more elusive. For example, in 1958 Mexico and Guatemala tussled over fishing rights.[37] The conflict occurred at a time when Latin American states were negotiating local rules governing the law of the sea, and Mexican shrimp boats were encroaching on Guatemalan fishing waters.[38] The dispute was over access to shrimp grounds. At its peak, Mexico sent over 200 boats into the disputed waters to protest Guatemala's attempt to restrict what were previously considered international waters. Guatemalan planes intimidated the boats, leading to three deaths before the states negotiated a settlement a month later. That same dispute would likely play out very differently today in the context of UNCLOS, which both Mexico and Guatemala have ratified. First, the rules outlined by UNCLOS could provide a focal point for identifying the location of mutual international boundaries. Additionally, the disputants could turn to specific international tribunals to adjudicate their disputes, rather than resorting to militarized threats. In the next section, we turn to a recent historical example to see how such institutions work in practice.

Romania-Ukraine Maritime Boundary Dispute

To illustrate the potential for such institutional management, let us take a look at the resolution of a maritime boundary dispute between Romania and Ukraine. After the breakup of the Soviet Union, Romania gained a new neighbor in the form of an independent Ukraine. The two countries needed to settle the boundary between their territorial waters and EEZs in the Black Sea. The location of this maritime boundary was an open question, as Romania and the Soviet Union had not been able to reach an agreement over this issue after two decades of negotiations during the Cold War. The stakes of the issue increased with the discovery of estimated reserves of 100 billion cubic meters of natural gas and more than 10 million tons of oil in a disputed area of the continental shelf.[39]

In 1997, Romania and Ukraine signed the Treaty on Relations of Cooperation and Good-Neighborliness. As part of this treaty, the two countries agreed to

---

37. Brecher and Wilkenfeld 1997.

38. Suman 1981.

39. BBC News 2009.

negotiate a mutual border regime, including the boundary between their EEZs in the Black Sea. They also stipulated that if they were unable to come to an agreement on these issues on their own, then either of the states could submit the claim for adjudication by the ICJ. In 2003, Romania and Ukraine signed a border treaty that finalized both the land border and the boundary between their territorial seas. However, after twenty-four rounds of negotiation over a seven-year period, they were not able to come to an agreement on the EEZ boundary.[40]

At the heart of the dispute, Romania and Ukraine disagreed on the proper way to draw the maritime boundary under international law. The primary source of this disagreement was the status of Serpents' Island,[41] a small islet (0.066 square miles) located 22 miles east of the mouth of the Danube River. In 1948, the Soviet Union occupied the islet, which had been previously controlled by Romania. Upon independence, Ukraine inherited control of the island from the Soviet Union.[42]

Romanian and Ukrainian officials had different interpretations of the role that Serpents' Island should play in the delimitation of the countries' maritime boundaries. Under UNCLOS, this role depends upon whether a landmass is a legally an island or a rock. Romania claimed that Serpents' Island was merely a rock and thus should be ignored when determining the extent of each state's EEZ. On the other hand, Ukraine argued that it was a proper island that was entitled to generate an EEZ like any other piece of territory. Due to these differing interpretations of the status of Serpents' Island, Ukraine claimed that the maritime boundary should be significantly farther south and west than the line proposed by Romania.[43]

To adjudicate the claims, Romania submitted the case to the ICJ in 2004. Five years later, the Court laid down its ruling. Sidestepping the question of the status of Serpents' Island, the Court ruled that the coastline of the island was too short to be consequential in determining the size of the states' EEZs. After taking into account Ukraine's 12-nautical-mile territorial sea around Serpents' Island, the Court drew a maritime boundary equidistant from the two countries' coasts.[44] The Court's boundary lay between those originally proposed by Romania and Ukraine. In the end, Romania gained control of 80 percent of the

---

40. Ivan 2007; Kruglashov 2011.

41. This island is also known in English as Snake Island.

42. Ivan 2007; Kruglashov 2011.

43. Lathrop 2009.

44. Ibid.

disputed maritime area, while Ukraine maintained the rest. Both countries welcomed the compromise decision and agreed to comply with the ruling.[45]

The case of Romania and Ukraine nicely illustrates how institutions can help states peacefully resolve property rights disputes. First, the provisions of UNCLOS provided rules that the two states could follow when drawing their mutual maritime boundary in the Black Sea. Second, when disagreements arose about how to apply these rules, the states turned to the ICJ to clarify this legal uncertainty and determine where the boundary should be. Through these rules and procedures, international institutions helped Romania and Ukraine reach a consensus on this critical issue when they were unable to do so on their own. Moreover, by clarifying the property rights of these two states, the institutional settlement opened the way for joint economic gains by allowing Romania and Ukraine to potentially move forward with oil and gas production in the area.

Summary

In this chapter, we have highlighted some key features of what we consider to be normal politics in the global economy. In the context of NEE, prices are set by supply and demand conditions in the market. Thus, buyers and sellers can be seen as price takers. Without an ability to affect prices, political actors in this environment largely focus their efforts on coordinating or manipulating the effects of the market price. In this way, the market sets the agenda for political activity in the NEE.

To help achieve efficient outcomes in the global political economy, states have created a variety of institutions that promote coordination and cooperation through the reduction of transaction costs and the creation and maintenance of property rights. At the global level, territorial borders and maritime boundaries delineate international property rights among states. When disagreements over the sovereign control of territorial and maritime resources arise, international institutions can provide an effective mechanism for resolving these property rights disputes. In particular, these institutions can provide rules for drawing boundaries to allocate property rights and procedures for reaching durable settlements.

The picture we have drawn in this chapter presents a global political economy in which markets are perfectly efficient and institutions are always effective. This is a good first approximation of the normal state of affairs and a useful starting point for understanding the nature of international relations. However, at times this regular order can break down and lead to inefficient and even violent

---

45. BBC News 2009.

outcomes. In this book, we are particularly interested in exploring two of these potential disruptions to NEE: efforts by actors to gain price-setting abilities in markets and the inadequacy of institutions to resolve some property rights disputes.

First, global actors sometimes have opportunities to take actions that would give them the ability to set prices. This price-setting ability would be valuable because it would provide the price setter with the potential to extract economic rents and gain political leverage. However, such a structural shift from perfect to imperfect competition would be inefficient for the market as a whole. Achieving such a price-setting capability requires a significant increase in market power, so we call these situations *market power opportunities*. Since a state will often need to expand its sovereign control over territorial or maritime resources in order to achieve a sufficient shift in market power, these market power opportunities can create or exacerbate international disputes over property rights, which provides an additional threat to NEE environments.

Second, institutions may not always be able to help states resolve property rights disputes, particularly when market power opportunities are present. In fact, as we will see, some of the same factors that make institutions effective—focal points and reputation costs—can sometimes deter states from pursuing institutional solutions to property rights disputes. If institutional rules and procedures would take a market power opportunity off the table or preserve the price-setting capability of a rival, states may prefer to pursue alternative routes to achieve their market power goals. In these situations, states may opt to strategically delay a resolution of the property rights dispute or even turn to war. Understanding when and why this happens is our task in the next chapter.

3

Market Power, War, and Strategic Delay

IN THE PREVIOUS chapter, we examined the role of institutions in promoting cooperation in the context of normal economic exchange. In this environment of perfect competition, buyers and sellers are price takers. However, not all markets operate in this typical fashion. In imperfect competition market structures, a small number of buyers or sellers have the ability to influence prices. The extreme cases of imperfect competition are monopoly and monopsony, in which one seller or buyer, respectively, dominates the market and can set prices. Similarly, under oligopolistic competition, a small number of sellers can potentially collude on higher market prices. Such an ability to set prices can be valuable, as these market imperfections provide opportunities to extract economic rents. Moreover, this market power can provide price setters with leverage over buyers and other sellers when bargaining over other issues.

Given the potential economic and political benefits that they can accrue, states face significant temptations to seize opportunities to give their firms the ability to set prices in global and regional markets. Of course, such opportunities do not present themselves every day. Achieving a price-setting capability requires a significant increase in market power. To achieve this, a state may need to gain sovereign control of a significant share of the supply of a good or seize substantial control of supply routes. This can lead to pressure to shift territorial and maritime boundaries to bring these resources under a state's jurisdiction. Thus, at their heart, these market power opportunities involve international disputes over property rights.

In this chapter, we establish an argument for why property rights disputes that contain market power opportunities are particularly difficult for institutions and states to navigate, leading to non-cooperative outcomes such as strategic delay and even the onset of war. To articulate this causal process, we develop a theoretical model of market power politics. In this model, we identify three key factors that can influence state behavior in property rights disputes: market power, economic interdependence, and international institutions. Opportunities to gain market power can motivate states to engage in expansionist behavior, while economic interdependence and institutions can potentially constrain states from using force to achieve their expansionist goals. The configuration of these factors in a given situation will thus influence whether states engaged in a property rights dispute choose to pursue strategies of war, delay, or peaceful settlement.

To develop this model of market power politics, we work through three key pieces of the puzzle. First, we identify when disputes over property rights become disputes over market power opportunities, thereby ceasing to be primarily concerned with the efficiency of economic exchange. In such situations, market power opportunities can motivate states to abandon the infrastructure of international institutions and seek more aggressive strategies to protect or capture new territory. Second, we examine when states are constrained by their economic ties such that they are unable to take advantage of market power opportunities. We are particularly interested in identifying situations where states would otherwise use more aggressive strategies but are prevented from doing so by their overall economic dependence on another state. Third, we reconsider the role that institutions play in property rights disputes. In the previous chapter we explained how institutions can help states effectively resolve disputes. However, in the presence of territorially based market power opportunities, the same factors that make these institutions so effective can sometimes deter states from pursuing institutional solutions in the first place.

When considered in tandem, the set of constraints created by international institutions and economic interdependence influences the strategies of states faced with market power motivations. If economic interdependence is low and institutions are not equipped to address the issues at hand, states will be less constrained and more likely to turn to violence to enhance their market power or to prevent a rival from doing so. At the other extreme, when high levels of interdependence make exiting economic relationships costly and institutions can provide a mutually beneficial agreement, states will be willing to pursue a peaceful dispute settlement.

Finally, in the intermediate case where disputants are constrained by economic interdependence, but institutions are ill-equipped to solve the problems that arise from market power opportunities, states pursue a strategy of delay until such time that they are able to resolve property rights disputes in their favor.

Motive and Constraint: A Model of Market Power Politics

In the previous chapter, we established that states are naturally inclined to resolve property rights disputes in pursuit of efficiency and development of economic exchange. While there may be important non-economic hurdles to resolving territorial conflict, in a normal economic exchange environment, the economic pressures point toward resolution. In addition, previous work on economic interdependence often points to the pacific effects of interdependence, and international institutions have emerged as an important tool in the post–World War II era to facilitate nonviolent conflict resolution.[1] All of this begs the question of why we think these ingredients can sometimes mix together in a markedly different fashion to produce political outcomes of war or delay that generate clear costs and inefficiencies for all parties involved. In other words, what are the economic motivations that can push states to set aside property rights management mechanisms developed by our international institutions, and what might constrain these states from using force even though these motivations are in place?

To understand the emergence of war or delay in these disputes, we need to first understand why states would be motivated to engage in non-cooperative behavior. There can be many such motivations, including distrust, enmity, and domestic political challenges. Perhaps former violence between states poisons the well, so to speak, preventing states from trusting one another to abide by the rules that institutions provide. Alternatively, domestic challenges to political survival could drive leaders to abandon cooperation in an attempt to appease supporting factions or to rally the population's support. Yet, institutions are designed to anticipate these issues and ameliorate them. Indeed, the history of the European Union is rich with gradual modifications to overcome the divisions that plagued post–World War II Europe. As such, we focus on states' motivations that are unmitigated by institutions and thus are incompatible with institutional design.

One such motive is economic competition. There are multiple ways in which competition in the sphere could lead to conflict between states. Chatagnier and Kavakli provide a useful analysis of how competition can trigger violence regardless of market structure.[2] They demonstrate that increased similarity in exports is associated with increased competition over import targets. Similarly, states that export similar goods may compete over the same inputs, triggering a scramble for raw resources. We focus our attention on economic competition that arises

---

1. Russett and Oneal 2001.

2. Chatagnier and Kavaklı 2015.

between states that want to change or preserve the structure of a market. In these cases, conflict emerges because states desire to give their firms sufficient market power to be able to set prices for an important good. We call a situation in which a state can take steps to give its firms price-setting capabilities a *market power opportunity*.

Market power opportunities can provide a motivation for states to escalate property rights disputes. Markets for many hard commodities, like hydrocarbons and rare earth elements, are intrinsically linked to territorial control. Producers in these markets require access to reserves of resources. If a state's firms were to gain price-setting capabilities in a valuable commodity market, the state could enjoy several benefits, including additional revenue from economic rents, political leverage in international bargaining, and increased stability. To gain or maintain market power in these commodity markets, states may have incentives to expand their territorial reach to secure additional resources, potentially through the use of force. At the same time, rival states may be willing to fight to prevent an expansionist state from securing a privileged market power position.

Of course, market power opportunities are not the only motivations for the violent escalation of property rights disputes. Existing research on territory and war identifies a variety of mechanisms that can produce such conflicts, including disputed homelands, historical claims to territory, and even resource-driven disputes. Our intent is not to supplant this large and valuable body of research. Instead, we intend to highlight the role of market power motivations in limiting institutionally managed cooperation. Our focus has dual motivations. For one, the literature has paid relatively little attention to the role of economic competition in these disputes. Moreover, as we will see, market power dynamics play a critical role in some of the most pernicious territorial and maritime disputes in the world today.

While states may have motives to achieve market power through violence, they are often constrained from doing so. Two forms of international constraints that shape decisions to pursue market power opportunities are economic interdependence and international institutions. Economic interdependence provides a constraint against violence, discouraging at least one side of the dispute from incurring the economic costs that would result from an escalation of conflict. Institutions provide mechanisms that can potentially help states peacefully resolve their property rights disputes. Clearly this is not an exhaustive list of constraints on violence. Most obviously, we leave out domestic sources of constraint, although both interdependence and institutions have domestic linkages. Domestic sources of constraint are not unimportant but tend to be varied in complex ways that require their own focus. As such, we take a partial equilibrium approach in this book and keep our attention attuned to the interstate level of analysis. These two sources of constraint that we consider are particularly interesting sources of friction that can be found at the interstate level of analysis.

How do these puzzle pieces fit together to explain when property rights disputes are prone to violence and strategic delay? Figure 3.1 provides a diagram of the overall process in our model of market power politics. Whereas market power opportunities provide motivation for aggression and the abandonment of normal dispute management processes, economic interdependence and international institutions serve as constraints that prevent conflict from breaking out between competing states. We consider three possible outcomes of these property rights disputes: peaceful dispute settlement, strategic delay, and war. In international situations where market power opportunities motivate states to abandon normal dispute-resolution mechanisms, sometimes international institutions are still able to facilitate agreements that enable states to avoid conflict. When economic interdependence conditions constrain at least one motivated state in these situations and the potential institutional settlements are not acceptable, we expect to see that state to attempt to strategically delay the resolution to their dispute. When international institutions are unable to facilitate an agreement and economic interdependence does not sufficiently constrain states from conflict, we expect an increased likelihood of armed conflict.

Thus, the emergence of violence or strategic delay in property rights disputes is a function of both the motives of states and the constraints that they face. In the discussion that follows, we more fully articulate our theoretical model of market power politics. First, we explain how market power opportunities provide a motivation for interstate conflict. We then consider the mechanisms by which the

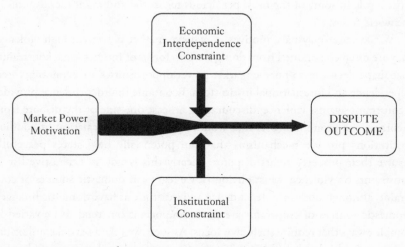

FIGURE 3.1. Motive and constraint in property rights disputes.

constraints provided by economic interdependence and institutions can reduce the chances of violent escalation and increase the likelihood of peaceful settlement. At that point, we will be able to put these pieces together to explain when these motives and constraints will lead to war and strategic delay in property rights disputes.

Motivated to War: How Market Power Can Lead to Conflict

Sometimes the economic goals of firms and their states are not based on efficiency or coordination. Instead, their goal may be to corner the market, which is a colloquial way to refer to establishing market power. Unlike the normal economic exchange that we discussed in the previous chapter, this kind of economic competition smacks of rivalry and zero-sum games and can have severe political consequences. Here we define and contextualize the concept of market power in order to help the reader differentiate between market power opportunities and efficiency-oriented exchanges within competitive markets. We then explore how market power competition can motivate states to pursue aggressive foreign policies, including the use of military force.

The Benefits of Market Power

Market power is the ability to generate prices that diverge from what would result from a fully clearing competitive market. In the last chapter, we discussed how normal economic exchange typically operates in markets that approximate perfect competition. In a normal economic exchange environment, both buyers and sellers are price takers. If a buyer or seller is able to gain sufficient market power, however, it can become a *price setter*. When this happens, the market transitions from perfect competition to imperfect competition. In this way, the emergence of price-setting capabilities is connected to the structure of markets.

While there are various forms of markets of imperfect competition, price-setting capabilities generally emerge in situations of market concentration when a small number of producers or consumers dominate the market for a good. If a firm controls a sufficient supply of a good, it can have the ability to set prices above what we would observe in a competitive market. This is known as *monopoly power*. Similarly, if a consumer sufficiently dominates a market, it may be able to demand lower prices. This is known as *monopsony power*. Empirically, monopoly and monopsony power are rarely observed in the absolute, and market power need not be defined as a complete journey to monopoly or monopsony. To achieve market power, a firm must be able to affect the price of the good or service that is being bought or sold. This, in turn, can lead to deviations from prices that would be set by supply and demand in a competitive market.

While it may be inefficient for the market as a whole, the ability to set prices can be very valuable for individual firms. When a producer can charge prices higher than the marginal costs of a good, it can extract profits beyond what a competitive equilibrium would predict. Likewise, a firm with monopsony power can increase its profits by paying below-market prices. The extra profits a firm receives from setting prices in its favor are referred to as *rents*. Given the value of these potential rents, firms have incentives to invest significant resources to gain the ability to set prices and to protect any market power that they accumulate. At the same time, other firms have incentives to block these efforts. This competition between firms to capture rents, which is known as *rent seeking*, has been the focus of a long-standing literature in economics and political economy.[3]

While much attention has been placed on the value of market power to firms, states can also benefit when their firms gain the ability to affect prices. Most directly, the economic rents that firms can accrue with their market power provide a potential source of revenue for the state. States could capture a share of these profits directly (in the case of state-owned firms) or through taxation. These rents will be especially valuable when firms have price-setting capabilities for a key good in the state's economy, such as its primary export. For example, a petrostate that largely depends on the production and export of crude oil to fill its government's coffers would greatly benefit economically if its firms could set prices and capture rents in oil markets.

In addition to economic benefits, market power can also provide both domestic and international political benefits to state leaders. At the domestic level, market power can potentially provide political stability for regimes. This may especially benefit rentier states, which are regimes that largely rely upon rents from commodity exports.[4] Autocratic regimes in rentier states stay in power by using revenue from these rents to either co-opt or repress domestic actors.[5] Since its earnings would be less dependent upon the actions of other global market actors, a rentier state with market power in a key commodity market may have a greater ability to generate a larger and more stable flow of these rents to ensure its survival. Leaders of rentier states with access to greater revenue streams may also be in a better position to avoid domestic political conflict. Evidence indicates

---

3. Krueger 1974; Posner 1975; Tollison 1982; Tullock 1967.

4. Beblawi and Luciani 1987.

5. Ross 2001; Wright, Frantz, and Geddes 2013.

that oil-dependent states with greater oil wealth per capita are less likely to experience civil wars than less wealthy oil producers.[6] These domestic political benefits of market power are not necessarily limited to autocratic regimes. Most notably, democratic governments in Botswana have used the rents gained from its market power in the diamond market to provide significant public goods and transfer payment that have helped them ensure electoral success.[7]

Market power can also enhance a state's position internationally. In particular, states can use their firms' market power as leverage to influence actors in other states. When its firms have price-setting capabilities over a key good, a state could use the credible threat of raising prices to extract concessions when bargaining with another state. Similarly, the state could reward other states who pursue favorable foreign policies with lower prices. For example, in a case that we will explore in more detail in Chapter 6, Russia has frequently used its leverage over gas prices in former Soviet states to privilege foreign leaders who in turn support ties to Russia rather than embracing the European Union. In this way, Russia and its natural gas firm, Gazprom, exchange profits for influence. Thus, market power can be translated into a form of coercive power in international relations. These political benefits can substantially enhance the value of market power to states beyond the pure economic gains from the capture of rents.

Market Power Competition

Given the potential economic and political benefits, states can have strong incentives to take steps to increase their firms' market power. As we noted earlier, we call a situation in which a state can take action to provide its firms price-setting capabilities a market power opportunity. At the same time, rival states will be motivated to prevent states from capitalizing on market power opportunities. This can create a competitive environment between states similar to the dynamics between rent-seeking firms.[8] Like situations in which firms are willing to expend significant resources to gain monopoly power or prevent other firms from doing so, market power competition between states can create economic inefficiencies and potentially lead to deviations from perfectly competitive markets. However, unlike competition between firms, this economic competition between states presents an additional danger, as it can sometimes lead to armed conflict.

---

6. Basedau and Lay 2009.

7. Dunning 2008.

8. Tullock 1967.

To conceptualize this environment of market power competition, imagine two governments as they seek to optimize the position of their industries, whether in an attempt to expand market share or to preserve it. To simplify things, assume that only one state has revisionist goals, while the other prefers the status quo. This could be a situation in which one state desires to increase its firms' market power, while a rival wants to prevent this from happening. Alternatively, it could be a situation in which one state's firms already enjoy price-setting capabilities and another state desires to take steps to reduce the former's market power. In either case, one state wants to change its market position, and the other wants to keep things the way they are. This kind of bargaining goes on constantly in international politics, often via semi-structured environments such as GATT/WTO negotiations or free trade agreements. Often these negotiations fully resolve the issue, but we are interested in issues that arise from competition that are not successfully resolved in these normal bargaining environments.

While it may be individually rational for states to engage in market power competition, this behavior can lead to two forms of inefficiency when compared to the normal economic exchange environments we described in Chapter 2. First, if a state's firms are successful in gaining or maintaining price-setting capabilities, this competition can lead to the establishment of markets of imperfect competition. Since prices in these market structures—which include monopoly, monopsony, and oligopoly—diverge from the market-clearing price, they are typically less efficient than perfectly competitive markets. Second, regardless of whether an imperfectly competitive market emerges, market power competition is inefficient because states engage in costly activities to pursue market power or prevent others from gaining it. Like war, this competitive behavior is *ex post* inefficient because the actors could have agreed to whatever market structure emerges from this competition without investing time and resources in competitive behavior. In this way, market power competition creates opportunity costs, as states could alternatively be putting their effort into more productive activities.

We should emphasize that this competition over market power does not necessarily entail the threat or use of force. For example, in the post–World War II economy, Japan developed a reputation for using government incentives to subsidize so-called sunrise industries, such as consumer electronics. The goal was to effectively drop the price of the Japanese goods below the market price and, more importantly, below competitors' prices. Sustaining these low prices required persistent government subsidization, but the long-run goal was to drive competition out of the market. The process, also known as dumping, was designed to affect market position. Once Japanese firms drove out competitors and claimed a large portion of the export market, these firms would begin to enjoy price-setting power. This would in turn allow them (in the short run, at least) to increase prices and extract rents from consumers. In the case of consumer televisions, the strategy partially worked. In the beginning of the second half of the twentieth century, dozens of American firms produced televisions. Four decades later, Zenith was the last remaining firm to assemble or produce televisions in the United States.[9] The Japanese attempt to control the television market was not a complete success, however, due to emerging producers in South Korea and China with government subsidies of their own.

Japan's efforts to dominate the television market did not translate into armed conflict, nor did they ever threaten to. Market power competition is not synonymous with violence; tension and violence are very different political outcomes. However, these tensions can sometimes escalate. In Part II of the book, we will explore several empirical cases in which states have turned to the use of force or gray zone tactics to pursue their market power objectives. Like the Japanese subsidization of consumer electronics, the goal of these states is to set the price of an important good. However, they have turned to very different foreign policy strategies to achieve this goal. This raises the question of when market power opportunities become important enough for states to fight over.

Criteria for Market Power Motivation

A state will be more motivated to take aggressive actions to pursue a market power opportunity when the value of price-setting abilities in a given market is high. As we outlined earlier, a state can benefit both economically and politically when its firms have the ability to set prices in international markets. On the economic side, the rents gained by firms with market power can provide a revenue source for the state. Additionally, market power can provide states with increases in political stability at the domestic level and greater bargaining leverage on the international stage. The size of these benefits is in part a function of the salience of the good, the market structure, and the nature of state-firm relations. Let us consider each of these factors in turn.

While perhaps it goes without saying, price-setting capabilities are more valuable for goods that are more salient to producers and consumers. If a good is a key export or makes up a significant portion of a state's economy, then the size of the potential rents available to the state will be higher. Additionally, given their importance, states will likely put a premium on providing stability in these

---

9. Since that time, a few firms have resumed American production, but the overall number of units produced is estimated to be a tiny fraction of American consumption. For our purposes, it is accurate to state that there have not been significant American producers in this market for over twenty years.

markets. Thus, petrostates that are heavily dependent upon the sale of oil, for example, would greatly benefit from the increased and stable stream of revenue that would come with the ability to shape prices in the oil market. At the same time, the political leverage derived from market power will be greater when prices for the good in question are important to other states. For example, given their critical importance to consumers, countries are generally very sensitive to increases in energy costs. This would give states with market power in oil and gas markets a greater ability to use price setting as leverage when bargaining with consumer states.

Market structure is another important factor in understanding when economic competition can lead to conflict between states. If global markets are competitive, any disruption of trade between two states will simply lead both to alternative markets.[10] When states operate within competitive markets, we can expect them to face some costs when adjusting to trade disruptions but no serious hurdles to maintaining the flow of goods and services. While these costs may be real, the existence of alternative economic partners fundamentally limits the ability of states to leverage economic ties for political gain. As Morrow points out, "[if] economic actors arrange their trading activities to account for the possibility of political disruptions of trade, then states have already paid the costs of conflict in trade before a dispute occurs."[11] Thus, when markets are structured such that no actor is a price setter, changes in competitiveness are unlikely to trigger conflict.

On the other hand, if markets are not operating in competitive fashion, market power can lead to more serious adaptation costs, which in turn can influence political conflict.[12] This can make market power opportunities more valuable for states who gain the ability to set prices and more threatening to others. For example, if there are a small number of producers in a market, consumers will find it more costly to adjust to shifts in market power. In the extreme, if a shift in market power leads to a monopoly provider of a good, buyers will have no alternative sellers to turn to. Adaptation costs will also be more costly when existing technology or infrastructure makes it difficult to shift between producers, such as in the pipeline natural gas markets that we will explore in more detail in Chapter 6. Finally, costs to consumers of adapting to shifts in market power will be greater when substitute goods are not available or would be costly to pursue. A corresponding set of adaptation costs for producers could arise in markets where a consumer

---

10. Gowa 1995, 520.

11. Morrow 1999, 48.

12. Crescenzi 2005.

has monopsony power. Thus, states are likely to have greater incentives to take aggressive steps to pursue or prevent shifts in market power in markets with a small number of producers or consumers.

The value to a state of a market power opportunity can also depend upon the structural relationship between the state and its firms. States that own or have significant control over their firms in a market will be better able to reap the benefits of market power. In these cases, the state will be in a more advantageous position to capture rents from price-setting firms. When a state owns a firm, it can more easily transfer profits earned by the firm directly to the government's coffers. States will also likely be able to more easily tax firms over which they have significant control. Additionally, such governments could have more influence over price setting, which would allow them to more effectively use the firms' market power as political leverage in their dealings with other states.

In more liberalized economies, however, the story is more complicated. While their firms will still desire price-setting capabilities, such states may be less able to enjoy the benefits of market power. These governments cannot direct access rents captured by price-setting firms, nor can they control the prices set by these firms. Thus, they are less readily able to generate revenue or political leverage from market power. This raises the question of how firms in these states could motivate their governments to take actions for the firms' benefit. Firms would need to find ways to lobby states to pursue market power opportunities, while governments would want to guarantee themselves a share of the benefits, perhaps through taxation of the profits that firms gain from their market power. Outlining this process in the domestic political economy—which will largely depend upon the nature of domestic political institutions and the makeup of the government's winning coalition—lies outside the scope of this book.[13] But regardless, we expect that leaders of states with liberalized economies will be less motivated to pursue escalatory foreign policies in pursuit of market power.

Putting all this together, we can now identify a set of criteria to determine when market power opportunities become important enough to fight over. First, these opportunities involve a key good for the state's economy. For states dependent upon rents from the primary commodity exports, that good could be a commodity they produce. For other states, the key good should make up a large enough portion of the economy to affect the overall health of the state's

---

13. This is not to say that the question is unimportant. Quite the contrary, this is an important expansion of our argument that will allow scholars to better understand how market power opportunities influence a heterogeneous set of institutions. But we need to first identify the basic mechanisms that shape market power politics before achieving a more complex understanding that melds domestic political economy with interstate market power politics.

economy if market prices change significantly. Second, shocks to the prices of key goods must be costly in terms of adaptation by other players in the market. Cornering a market is not a source of power if that market can be set aside by its participants. These motivations will likely arise in markets with a limited number of producers or consumers. Finally, all else being equal, we expect states that own or have significant control over their firms in a market will be more likely to take aggressive steps to gain or preserve market power than states with more liberalized economies.

Thus, when a small number of producers or consumers can influence prices and the economic activity is sufficiently salient, states have greater incentives to actively respond to changes or threats of changes to the status quo. States that enjoy an advantage in an important market may be willing to fight to keep it, and states that see the potential to obtain an advantage over competitors may be willing to fight to get it. In essence, such changes in competitiveness can be thought of as changes in state power. Not all changes are equal, of course, but if the change in power is large enough, then it may motivate a state to alter its political behavior. To further explore these dynamics, we want to focus our attention on a set of markets in which we expect that this type of behavior is likely to occur. In particular, we will examine how market power competition in hard commodity markets can lead to the escalation of property rights disputes between states.

Hard Commodity Markets and Property Rights Disputes

Our aim in developing a model of market power politics is to understand factors that can lead states to abandon institutional resolutions to property rights regimes and instead turn to violence or gray zone tactics. Thus, while states can take steps to secure market power for firms in a wide variety of markets, we want to focus our attention on market power opportunities that arise around goods that are intrinsically linked to territorial or maritime control. Perhaps the clearest category of such goods are hard commodities, which are natural resources that must be mined or extracted. Examples of hard commodities include energy resources (e.g., oil, natural gas, coal) and metals (e.g., gold, silver, rare earth elements). Since the production of hard commodities requires access to reserves of these resources, market power in these markets is in part a function of who holds the property rights over the land or seabed where these reserves are found. We expect that market power opportunities in hard commodity markets are likely to be valuable to states and to have the potential to motivate them to pursue expansionist foreign policies.

Markets for many hard commodities have the potential to meet the two criteria for market power motivation that we outlined earlier. First, these commodities are often salient for both producers and consumers. Notably, many states are highly dependent upon commodity exports, and the profits from this trade can be an important source of government revenue. Thus, the ability for a state's firms to set prices in these markets could provide an increased and steadier flow of revenue that would help the political survival of the state's leaders. At the same time, many of these resources—including the examples of oil, gas, and rare earth elements that we will explore in Part II of the book—are also of critical importance to consumers around the world.

Additionally, the structure of these markets can provide valuable market power opportunities. Both the geographic distribution of resources and infrastructure requirements for extraction and transportation can limit the entry into these markets. Given this, there tend to be few producers in these markets. This makes deviations from competitive markets more likely, as it is easier for an individual producer to gain control of a significant share of the available resources to set prices. These factors also reduce opportunities for consumers to turn to alternative producers, which creates high adaptation costs. This increases the potential rents and bargaining leverage that could come with the ability to set prices in these markets.

Given these anticipated benefits, market power competition in hard commodity markets can motivate states to expand their territorial reach. Since production of these commodities typically requires a state to have access to resource reserves, market power is often a function of territorial control of these reserves. If a state wants to increase its market share or prevent another state from doing so, it may need to expand its territorial control. Additionally, in some cases, states can augment their market power by controlling supply routes for the good. This can lead to pressure to shift territorial and maritime boundaries to bring these resources or transportation routes under a state's jurisdiction. In this way, market power competition can create or exacerbate property rights disputes between states.

In the previous chapter, we discussed how international institutions can help states resolve such property rights disputes. However, as we will further explore, states may not be able to achieve their market power goals through an institutional settlement. In these cases, states may turn to more aggressive strategies to achieve their expansionist goals. In the extreme end, states may attempt to seize territory through the use of force. Despite the costs of war, states may be willing to fight if the market power opportunity is sufficiently large. Seizing valuable territory that provides the ability to set prices in a key market can lead to a discontinuous shift in power, creating a commitment problem.[14] If states expect that war

---

14. Powell 2006.

will potentially provide them with a future advantage in market power that leads to long-term compounding rewards, they may be willing to forgo a settlement and turn to violence.[15]

Even in cases where states do not opt to go to war to achieve a market power opportunity, they may still be unwilling to make a settlement. In that case, they may choose to strategically delay a settlement of the property rights dispute. Specifically, if settling a property rights dispute removes the market power opportunity for a state, then the state has an interest in avoiding settlement despite any normal economic and political benefits that may ensue. At a minimum, delaying allows the state to wait and capitalize on any future change the international strategic environment that would open up the possibility of a settlement in line with its market power goals. Alternatively, the state could take a more proactive approach and use a period of delay as an opportunity to gradually expand its territorial reach through the use of gray zone tactics. In either case, states may resist the resolution of property rights disputes due to the expectation of an improved bargaining position in the future.

To this point, we have mainly focused on the willingness of states to pursue aggressive strategies to expand their territorial reach in order to seize a market power opportunity. However, some states may also have incentives to pursue strategies of violence or delay to prevent another state from establishing market power for its firms. If a state fears that a rival is likely to use force to take control of territory that would lead to a discontinuous shift in market power, it may be willing to take military action to preempt such a move. Additionally, a state may strategically delay settling a property rights dispute if it expects that a settlement would bolster the market power goals of a rival.

Moreover, we expect that states may be willing to turn to the use of force or strategic delay even if they do not have the potential to benefit from having market power themselves. Instead, these countries may be motivated to take these steps to prevent a market power opportunity in order to protect their citizens from paying high prices in an important market. Alternatively, they may be driven to prevent a rival from gaining the political leverage that can accompany market power. For these reasons, we may see a wider range of states beyond direct market competitors who are willing to engage in this preventive behavior. In particular, since these motivations are not connected to a state's ability to benefit from its firms' market power, we expect that states with liberalized economies will be more likely to engage in strategies of market power prevention than strategies of market power pursuit.

---

15. Garfinkel and Skaperdas 2000.

In sum, we expect that market power competition in hard commodity markets can sometimes motivate states to pursue aggressive strategies to expand their territorial reach or prevent other states from doing so. However, states with these motivations will not always turn to the use of force to achieve their market power goals. Instead, they may opt to pursue a strategy of delay or may even be willing to reach a settlement to their property rights dispute. To understand when states choose war, strategic delay, or dispute settlement in the face of market power motivations, we need to also consider the constraints that these states face. We now turn our attention to outlining two general categories of these constraints: economic interdependence and international institutions.

Economic Interdependence as a Constraint to War

Up to this point, we have focused on how market power competition can motivate states to fight. However, much of the study of economics and conflict has actually focused on the constraining effects of economic interaction on the use of military force. Our goal in this section is to identify the economic conditions that can act as a counterbalance or constraint against using force in pursuit of market power opportunities. It may at first seem illogical to argue that economic exchange can both induce and constrain violence. The key, however, to understanding how economics affects political decisions is to identify the conditions that encourage the pursuit of efficiency gains versus those that incentivize the pursuit of price-setting power. One key factor that pushes states in the former direction is economic interdependence. By increasing the costs of exiting economic relationships, interdependence discourages states from fighting and can potentially take the option to go to war off the table.

To understand how economic interdependence can constrain state behavior, we need to consider the difference between two types of costs that actors can face when exiting economic relationships: sensitivity costs and vulnerability costs.[16] *Sensitivity costs* are the inefficiency costs that economic firms and agents incur when changing economic partners in a normal economic environment. Even in markets that are functioning close to perfectly, it is reasonable to expect that economic ties are generally formed because they are the most efficient and they minimize transaction costs. Adaptation to economic disruption in this environment is still costly, but these costs are much lower and shorter in duration. As economic agents adapt to new partners, they need to invest in new contracts, new transaction-based institutions, supply chains, and other dimensions of economic

---

16. Baldwin 1980.

exchange. This adaptation tends to be temporary, with a short time-horizon. There may be small efficiency losses, in that we assume the original economic partner was in that position due to its competitive position in a normal market. The closer the market is to "perfect" in terms of alternative buyers and sellers, the smaller these efficiency losses will be.

For an example of these sensitivity costs, consider Volvo Corporation's supply chain management. According to Volvo's supply chain sustainability report, the corporation maintains approximately 3,500 preferred suppliers who deliver parts that are then used to manufacture and assemble Volvo vehicles.[17] This list is curated, and the curation process is driven by a combination of economic factors (e.g., quality, price, scalability) as well as policy factors (e.g., code of conduct standards, ISO safety certifications, environmental concerns). Whenever Volvo chooses to sever ties with one supplier, it has multiple alternative suppliers that adapt to the change in demand, and it incurs short-term costs while it establishes ties with a new supplier. These costs are sensitivity costs, and businesses avoid them when possible, but rarely do they threaten the overall health of the firm's economy.

Vulnerability costs, on the other hand, are more severe and require more difficult adaptation strategies. If, for example, Volvo had only one supplier of tires for its vehicles and based on economic and policy standards no other tire manufacturer produced tires that would be suitable for Volvo vehicles, then the loss of that economic partner would require much more severe adaptation. As such, vulnerability is driven by two dimensions: alternative economic partners and one's own reliance on the goods and services being exchanged.

When the number of alternative potential economic partners becomes constrained, the costs of exiting the economic relationship increase. The edge case involves zero alternative partners, which forces the firm or the state to choose between its current partner and forgoing the goods or services exchanged with that partner. Even in this edge-case scenario, the actual costs of economic exit depend on the salience of the economic good or service involved. Returning to the Volvo example, if Volvo found itself in a situation where it had only one supplier of tires, it would find it very difficult to stop buying tires altogether. If, however, Volvo wanted to introduce a new luxury item in its sedans such as OLED screens for rear passenger viewing, and only Samsung provided the screen technology, Volvo could opt to forgo the new addition rather than pay rents to Samsung. Thus, the ultimate cost of exit is jointly determined by substitutability and adaptation. Similarly, if a state has only one supplier of natural gas and that

---

17. Volvo 2018.

source of energy is essential to heating homes throughout the winter, that state would find it very difficult to refuse whatever price the exporting state demands.[18]

Thus, *economic interdependence* occurs when the risk of severing economic ties with a partner (economic exit) exceeds the expected value of pursuing demands that can lead to political violence.[19] This interdependence is contextual and is defined by the relative utility of the issue at stake as well as the costs of economic exit. The costs of economic exit are a function of the value of the economic exchange as well as the opportunities to obtain this exchange from other partners in the market.

There are two common mistakes made by those who study interdependence. First, the phrase is often misapplied to high levels of economic interaction between two states. But high levels of trade that are easily substituted from other buyers or sellers are not situations of interdependence. Instead, they are better characterized as situations involving sensitivity costs that are easily surmountable. Moreover, low levels of trade can still trigger interdependence if the goods or services being exchanged are sufficiently salient to one or both partners and there is no alternative supply or demand. The key characteristic that separates interdependence from interaction or interconnectedness is the costs of adaption that would be borne by one or both states in the event that economic exchange is severed. If these costs of adaptation are small or short-term and alternatives to the severed exchange exist, the economies can be thought of as interconnected, but not interdependent. Economic interdependence only exists when adaptation to economic exit is constrained.

The second mistake commonly made by scholars is to associate economic interdependence in concept or measurement strictly with the phenomenon of international trade. Trade is the exchange of goods and services across markets, and as such it represents only one dimension of economic ties. Economic interdependence can also occur as a result of capital flows, exchange rates, foreign direct investment, and external government debt. Moreover, large amounts of trade flows may not represent interdependence at all, particularly if they are primarily composed of intra-industry trade.[20]

---

18. While these joint conditions of substitutability and adaptation are intuitive, they introduce interesting challenges to data collection. These challenges, along with the challenges of identifying the set of potential market power opportunities in the global economy a priori, make testing our arguments difficult.

19. Baldwin 1980; Crescenzi 2005; Hirschman 1945; Wagner 1988.

20. Intra-industry trade is trade that does not result from specialization (and comparative advantage). Examples include Germany exporting Mercedes brand cars and importing Ford brand cars. Research indicates that intra-industry trade has markedly different effects on political conflict than trade resulting from specialization (Thies and Peterson 2015).

The type of economic exchange (trade, foreign direct investment, or debt) is less important than the costs of adapting to lost interconnectedness. Similarly, the amount of goods or money being exchanged is not necessarily telling. Certainly, all else being equal, increased interconnectedness suggests increased salience in the economic ties between states, but the source of constraint emerges from the costs of adaptation. Adaptation costs can be high even when the dimension of economic exchange seems small. In the 1980s, for example, South African exports to the United States never rose above 3 percent of total US exports, yet the United States was unwilling to properly exit this economic relationship in pro-test of South Africa's Apartheid government.[21] Despite the low overall amount of trade, the United States depended on key South African exports, including chromium, manganese, and platinum. The inability to find alternative sources for these minerals caused the United States to be economically dependent on South Africa and deterred the US from imposing impactful economic sanctions to sway South African policy.[22]

As such, it is useful to think of the constraining influence of economic inter-dependence in terms of adaptation and the costs of exit. When the costs of eco-nomic exit increase, so too does the constraint of economic interdependence on the ability to make demands and use violence. As the value of the issue at stake increases, however, we need to see higher levels of exit costs to realize these constraints. This interdependence need not be symmetrical. Asymmetrical inter-dependence provides the less constrained state with opportunities to seek politi-cal concessions from its partner. As long as those demands fall under the costs that would result from economic exit, the more dependent state will prefer to cooperate.

Economic Interdependence as a Constraint

Economic interdependence increases the costs and risks of using violence. When states are economically interdependent, at least one side cannot afford the adap-tation costs of exiting an economic relationship with the other state. The sources of these costs can be intrinsic (e.g., due to factor endowments or sectoral advan-tages) or political (e.g., strong domestic actors generate political costs associ-ated with economic exit). Exit costs also reflect the lack of secondary economic options, meaning the firms of one state cannot simply begin buying or selling their salient goods with other sellers and buyers in the market at a slightly higher

---

21. Barbieri 1995; Crescenzi 2005.

22. Crescenzi 2005.

cost. When adaptation is intensely constrained and economic exit would produce major political consequences for a state's leader, that leader will avoid policies that trigger such exit. Thus, if armed conflict would result in economic exit, states that are dependent on one another will face potentially destabilizing economic costs that may be sufficient to deter violence.

The dynamics of this constraint of economic interdependence on the use of force work differently depending on the particular symmetry of interdependence. Consider a property rights dispute between a challenger state that desires change and a target state that wants to maintain the status quo. When the target state cannot afford economic exit, the challenger state is able to make demands that improve its stake in the property rights dispute without fear of major economic repercussions. The size of these demands varies relative to the target's assessment of the value of those demands relative to the costs of exit. The higher the costs of exit, the more extensive the demands made by the challenger.

As we will explore in detail later in the book, this is the dynamic that currently exists between China and the Philippines, where China is the challenger and the Philippines is the target. Economic exit would trigger adaptation costs for both states and their firms, but China is in a far better position to deal with those costs than is the Philippines. This inability to live without the other prevents the Philippines from taking an active military stance against China to defend its disputed islands. Moreover, it also prevents the willingness of the Philippines to appeal to the United States as an ally to aid in this defense. As such, China is able to make small but steady progress in expanding its territorial reach in the South China Sea.

When the challenger is dependent on the target for economic ties, interdependence still constrains violence. In such cases, the target is able to resist the challenger's demands without fear of pushing the challenger to use violence. As long as the level of resistance triggers costs that are perceived by the challenger state to be less than the costs of economic exit, the challenger state will avoid escalated conflict and the exit that results. This dynamic can be illustrated by the current dispute between China and Japan over the Senkaku/Diaoyu Islands. In this case China (the challenger) is economically dependent on Japan's (the target) level of foreign direct investment.[23] Over the last half-century China has repeatedly made territorial demands on these islands, and each time these demands have been resisted effectively by Japan.

---

23. In this case, both sides are economically dependent on the other, making this a case of mutual or symmetrical interdependence. We discuss this more extensively in Chapter 7.

While the mechanics of the constraint vary with the symmetry of interdependence, in each case it is evident that these economic constraints have political consequences. With respect to the opportunities that emerge when states seek changes in property disputes in order to realize market opportunities, economic interdependence matters when the immediate costs of adaptation outweigh the short-term gains of conflict.[24] The Chinese cases also illustrate that while economic interdependence can remove the specter of war, it can also reduce the need to cooperate.[25] If one state is dependent upon another, it cannot credibly threaten to exit the economic relationship or escalate with political violence. Without such leverage, the dependent state would also have little ability to force its rival to participate in a dispute-settlement process if the rival does not want to do so. In such cases, the result may be strategic delay. Thus, the constraint of economic interdependence is not sufficient to guarantee the settlement of property rights disputes in the face of market power opportunities. To understand when states will be more likely to peacefully settle these disputes, we also need to consider the role of institutional constraints.

The Limits of Institutional Constraints

Like economic interdependence, international institutions can potentially constrain the motivations to engage in violence to seize market power. As we discussed in the previous chapter, a primary function of international institutions is to help states coordinate and cooperate more efficiently. In short, these institutions help states avoid suboptimal outcomes. In the context of property rights, institutions can both help states decide how to clearly allocate those rights and protect that allocation through the establishment of formally recognized international boundaries. They do this in two primary ways. First, institutions provide rules that guide states on how they should allocate property rights. Second, they provide procedures to help states reach settlements that are in line with the rules laid out.

Thus, at first glance, institutions would seem to be unambiguously helpful, and the development of strong institutions would seem to be a straightforward solution to conflicts that emerge from disputes over territorial and maritime control. Unfortunately, the story is not that simple. As it turns out, the presence of institutions can actually discourage dispute resolution in some situations. In fact,

---

24. State leaders also factor in the discounted expectations of future market power realization, but they do so relative to the strategies at hand (Copeland 1996).

25. Crescenzi 2005.

some of the same factors that make institutions such effective mechanisms of dispute resolution—focal points and reputation costs—can sometimes reduce the willingness of states to reach a settlement in the first place. Thus, the ability of institutions to constrain aggressive behavior in the face of market power opportunities will vary depending upon the strategic situation.

Institutional rules can help states coordinate on bargaining outcomes by providing focal points, but they also constrain the choices that states can make. This is not problematic when rules guide states to make settlements that are mutually beneficial. However, the strategic calculus changes when institutional rules require states to make choices that they view to be politically unacceptable. For example, consider the long-standing maritime dispute in the South China Sea. Any potential settlement that would be in line with the rules laid out by United Nations Convention on the Law of the Sea (UNCLOS) would likely require China to give up its claim to much of the South China Sea. In light of this, China has been reluctant to accept an institutionally prescribed outcome. At the same time, China would also prefer to avoid any reputation costs it would face by flouting international law, so it has not pressed a settlement that would be at odds with its obligations. Instead, China has allowed the dispute to continue unabated. In such situations where the only outcomes in line with institutional rules are politically unacceptable and the only politically acceptable outcomes are not in line with institutional rules, dispute resolution through institutions becomes difficult.

Similarly, states may have incentives to avoid institutional forums like international courts and arbitration panels. The effectiveness of these forums raises the stakes of using them. Because states must give up decision or process control to the institution, the use of these forums limits states' ability to structure any settlement over the issue. Since it is often difficult to renege on legally binding settlements, delegating such power to an international court or arbitration panel can be particularly risky. When faced with highly salient issues, such as the allocation of property rights, states may want to avoid the risks of an institutional solution. Instead, they may try to settle the issue on their own. However, states often cannot do this on their own, which is why these institutions were created in the first place.

Before moving forward, we want to emphasize that we are not claiming that institutions are always—or even often—counterproductive. Institutions can and do play an important role in promoting international cooperation. Instead, we argue that institutions developed to address a particular set of problems can sometimes have unintended consequences in international relations. Rules can provide focal points to facilitate coordination, but they can also unnecessarily constrain the range of settlements available to disputants. Similarly, potential reputation costs can increase the willingness of states to comply with institutional settlements, but they also raise the stakes of any agreement that is reached. Our goal is to understand the conditions under which institutions designed to resolve disputes may be less effective at constraining the aggressive behavior of states motivated to capture significant market power.

To examine the limits of institutions in the face of market power opportunities, we consider specific aspects of the institutional environment surrounding property rights disputes that can potentially inhibit successful dispute resolution. First, the costs involved in altering legally established territorial and maritime boundaries largely take renegotiation of boundary settlements off the table. This raises the stakes of any settlement that states reach to resolve property rights disputes and reduces the willingness of states to make concessions over territory. Second, given the institutional rules that constrain the set of potential boundary agreements allowed under international law, successful resolution of property rights disputes may require side payments. However, given the commitment problems that can arise in bargaining over strategic territory, the successful transfer of such side payments may not always be credible. This may hinder any chances for states to reach an agreement over the allocation of property rights. Most critically for the theory that we develop here, the potential for market power opportunities tends to exacerbate these challenges to the successful peaceful resolution of property rights disputes.

Inability to Renegotiate

In contemporary international relations, once territorial and maritime boundaries are settled under international law, they are difficult to change. Interstate disputes over property rights typically arise from competing legal claims to territory or maritime areas. By cooperating to settle the dispute either through a negotiated settlement or a legal ruling, states clarify their legal rights over the disputed territory. Once the issues of international law are settled, states can find it difficult to credibly make a claim for an alternative boundary. Unless one of the parties is able to demonstrate that international law was incorrectly applied, future legal bodies would likely reaffirm the settled boundary. States are also reluctant to reject borders established under international law if such a violation would generate a reputation cost that would hurt their bargaining position in other disputes.[26] Due to the ex ante cost of signing an agreement or the ex post costs of reneging, states that

---

26. Prorok and Huth 2015.

formally settle boundary disputes are also less likely to fight militarily over the formerly disputed territory.[27] Moreover, the emergence of a territorial integrity norm in the international system discourages the use of force to revise territorial boundaries.[28]

For these reasons, states may consider settlements that resolve boundary disputes to be largely non-renegotiable.[29] In the context of the politics of normal economic exchange, this stability of international boundaries is beneficial. By clarifying the allocation of property rights and jurisdictional control, a boundary acts as an institution that reduces uncertainty and promotes economic exchange. When economic actors expect that a boundary will remain unchanged well into the future, they will be more willing to pursue beneficial investment and trade opportunities.

While stable boundaries can promote efficient economic exchange in the long run, the inability to renegotiate can complicate bargaining attempts to resolve outstanding property rights disputes. In practice, states expect that any property rights settlement they reach will be long-lasting. This significantly raises the stakes for disputants. If a state ends up with an unfavorable settlement, it will either have to live with it or face the possibility of paying significant reputational costs for violating international law. Given these increased stakes, states will have an incentive to bargain harder in disputes of territorial control. This type of strategic bargaining environment often results in a lengthy stalemate over the issue at hand.[30]

While the inability to renegotiate boundary settlements always has the potential to complicate the settlement of property rights disputes, the magnitude of these challenges will vary from case to case. When the relative bargaining power between states is not likely to change over time, the inability to renegotiate may not have as significant an effect on the disputants' bargaining behavior. Typically, disputants will be willing to accept agreements that reflect their relative bargaining power. Thus, in situations where bargaining power does not change, a settlement that reflects the disputants' relative bargaining power will likely remain acceptable in the future. Since the disputants would have less incentive to renege on such an agreement in future periods, the inability to renegotiate will be less troublesome.

---

27. Schultz 2014.

28. Atzili 2011; Zacher 2001.

29. Carter and Goemans 2011.

30. Fearon 1998.

However, the potential for shifts in relative bargaining power could alter the strategic incentives of disputants. Suppose that there is a dispute between two states and that there will be a shift in power in favor of one of the states. With the possibility of renegotiation, the disputants could reach a settlement based upon their relative bargaining power before the shift. After the shift, they could then renegotiate the settlement with terms more favorable to the state that gained power. However, if there is no possibility of renegotiation, such a strategy is not possible. The states can either reach a settlement before the power shift, or delay and try to reach a settlement after the power shift. If a state anticipates a significant favorable shift in its bargaining power and is sufficiently patient, it may prefer to delay settlement of the dispute until it is in a more favorable bargaining position. In this case, rather than accepting a settlement on less favorable terms, the rising state would be willing to forgo the potential economic gains from resolving the property rights dispute in the near term to achieve a more favorable settlement in the future.

Market power opportunities can exacerbate the barriers to bargaining that arise from the inability to renegotiate property rights settlements. States that anticipate future market power opportunities will be particularly reluctant to settle property rights disputes. As we have noted, significant shifts in market power can lead to increased bargaining power. In this way, a state anticipating a market power opportunity is analogous to the rising state described in the previous paragraph. It will prefer to not reach a negotiated agreement when it is in a less advantageous bargaining position. Instead, it will have incentives to delay settlement until such time that it can use additional market power as leverage to achieve a more favorable allocation of property rights.

More importantly for our discussion here, the possibility of territorially based market power opportunities can heighten these incentives to delay settlement of property rights disputes. Since access to territorially based resources can be the source of market power, a settlement that results in a particular distribution of territory can affect the future bargaining power of states. For a state that hopes to gain price-setting capabilities, a non-negotiable settlement that cedes strategic resources to a rival state can take future market power opportunities off the table. If the current bargaining environment does not allow the state to gain access to such resources, it might prefer to delay settlement to keep its options open. Doing so would allow the state to take actions to put itself in a better bargaining position to reach a more favorable settlement at a later date. Thus, delaying a settlement leaves open the possibility of future market power opportunities.

Of course, the resolution of a property rights dispute requires agreement by all parties. Thus, states may also avoid a settlement if it would give a rival a significant increase in market power. In situations where states bargain over resources that provide bargaining power, commitment problems may arise because states fear that a rival will use its future advantage to demand an even greater share of the resources.[31] One might expect that the inability to renegotiate settlements would help alleviate these commitment problems. However, states with competing claims over mutual boundaries typically interact in a variety of economic and security arenas. Thus, an increase in market power can give a state increased future bargaining power over a wide range of issues beyond the division of territory. So even if a state cannot use its increased market power to gain additional territory, it would still be able to use its market power to gain advantages in other issue areas.

Credibility of Side Payments

While theories of territorial division often assume that disputants can choose from a continuous set of outcomes, structural factors often restrict the set of potential settlements to property rights disputes or privilege some outcomes over others. For example, geography may privilege some boundary settlements over others. Borders that align with preexisting sub-state administrative frontiers reduce transaction costs and are more stable over time.[32] Features such as rivers or mountain ranges can provide more defensible borders. Settlement patterns may make certain pieces of territory difficult to divide or may make boundaries that separate population groups' focal points.

As we noted in Chapter 2, institutional rules can also constrain the set of potential divisions of property rights. For example, settlements reached through arbitration or adjudication typically must be consistent with international law. States that have ratified UNCLOS are obligated to follow the principles of maritime delimitation outlined in the Convention. These rules provide focal points to help states coordinate on how to allocate property rights. However, limiting the available options can sometimes inhibit dispute resolution. Stable settlements generally need to reflect the relative bargaining power of the disputing states.[33] Bargaining power in property rights disputes can come from multiple sources, such as the strength of a state's legal claim, its military and economic capabilities, and domestic audience costs. If the set of potential settlements is restricted, there may not be a feasible division of property rights that adequately reflects the bargaining power of the states.

---

31. Fearon 1996; Powell 2006.

32. Carter and Goemans 2011, 2014.

33. Gent and Shannon 2014; Powell 1999.

When the set of available divisions of property rights does not reflect the relative power of the states, a stable settlement of the dispute may require side payments to bring the overall outcome in line with the balance of power between the disputants. For example, when Malaysia and Brunei settled their maritime boundary dispute off the coast of Borneo in 2009, Malaysia accepted that Brunei had sovereignty over the disputed area due to Brunei's strong legal claim. However, as a side payment, Brunei agreed to jointly develop the area commercially with its more powerful neighbor. Additionally, at the same time, the countries agreed to settle their long-standing territorial claims in the Limbang region largely in Malaysia's favor. Through the use of side payments and issue linkages, Malaysia and Brunei were able to peacefully resolve their property rights dispute in line with the rules of international law.[34]

However, at times, states may face problems credibly making the necessary side payments to reach a stable settlement because the allocation of property rights can alter the relative bargaining power of states, especially when one state achieves a significant increase in market power. If a state finds itself in a more advantageous position after gaining control of strategic territory or valuable resources, it will be less willing to make concessions at that time. This can potentially lead to credibility problems in two types of situations. First, commitment problems can hamper bilateral negotiations, as a state may be reluctant to sign on to an agreement if it believes that its counterpart will later renege on its obligation to provide side payments. Second, in the case of arbitration or adjudication, any agreement about side payments can only be made after a ruling has been made. Since a state that is advantaged by the ruling may no longer be willing to make side payments, it may be difficult to compensate "losers" in legal rulings. This increases the risk of legal dispute resolution, reducing the incentives for states to pursue this route.

In sum, due to factors such as the inability to renegotiate or credibly provide side payments, established institutional rules and procedures sometimes may not be effective in resolving property rights disputes when market power motivations are present. For states eyeing increased market power, institutional settlements may sometimes take potential price-setting opportunities off the table. In other situations, states may be concerned that an institutional settlement will preserve a rival's privileged position in a market. While these factors make dispute settlement more difficult, it does not mean that institutions are useless in the face of market power opportunities. When institutional rules and procedures produce settlements sufficiently beneficial to both sides, they can help resolve property

---

34. Roach 2014; Smith 2010.

rights disputes. However, these situations will be less common when market power opportunities are present than in normal economic environments.

Outcomes: Explaining War and Strategic Delay

We can now put the pieces of our theory together to provide an explanation for the outcomes that emerge in property rights disputes in the context of market power opportunities. In Chapter 2, we described how institutions can facilitate the peaceful settlement of property rights disputes in the environment of normal economic exchange. To be sure, successful institutional management is also possible when a state faces a market power opportunity. However, such a settlement is less likely due to the motives and constraints that disputants face in these situations. Instead, two other outcomes become increasingly likely: war and strategic delay. We now consider how the factors of market power, interdependence, and institutions interact to influence the likelihood of each of these outcomes.

War

As we have discussed, states presented with market power opportunities have motives to act aggressively to achieve a privileged economic position. In particular, states may be tempted to use military force to seize control of territory or gain access to resources that would provide a marked increase in market power. If the potential shift in market power is sufficiently large, this may create a commitment problem. Given the long-term compounding rewards that come with price-setting capabilities, an expansionist state may choose to forgo a settlement and instead turn to war to achieve its market power goals. Additionally, if a rival state anticipates such a move, it may opt to take military action to prevent such a shift in market power from happening. Of course, in many situations, factors such as economic interdependence and international institutions constrain states from using this military option. When these constraints are weak relative to the motivations to capture market power, states will be more likely to turn to violence.

Figure 3.2 illustrates the conditions under which we expect to see violent escalation of these property rights disputes. When states face market power motivations and are not constrained by economic interdependence or international institutions, they will be more likely to pursue war to achieve their goals. The presence of a market power opportunity motivates a state to engage in aggressive behavior to gain control of resources. Without a significant level of interdependence between the disputants, there are fewer economic costs of seizing these resources by force. Additionally, when institutions are not able to adequately

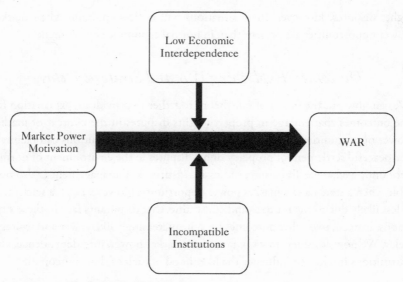

FIGURE 3.2. Conditions leading to war.

handle the issues surrounding market power, states will be less able to reach a peaceful settlement of the dispute. Thus, in the absence of strong economic or institutional constraints, armed conflict becomes an attractive option for states eyeing a significant increase in market power.

Of course, this is not a deterministic relationship. We do not expect that all states faced with market power opportunities will engage in war, just that they will be more likely to do so. Given the costs of armed conflict, if states can achieve their goals without the use of force, they will likely do so. In some cases, states may be able to gain access to desired territorial resources through nonviolent means. For example, Russia was able to seize control of Crimea in 2014 largely without the active use of military force. However, states rarely have the ability to pursue such a large-scale annexation in a relatively short amount of time through non-military means. Instead, nonviolent strategies to pursue market power opportunities are typically more gradual in nature. We now turn to the conditions under which we expect states to pursue such a delay strategy.

Strategic Delay

We often observe states pursuing nonviolent delay tactics in interstate disputes over property rights. Disputed claims over territorial and maritime boundaries often linger for decades without any clear resolution. In these cases, disputing states are unable or unwilling to settle the issue through either a bilateral settlement or a legal ruling by a third party. On the other hand, despite the fact that disagreement over territory is one of the most predominant causes of war, states often refrain from escalating boundary disputes to armed conflict. Instead, disputants opt to maintain their incompatible claims and perpetuate uncertainty about territorial and maritime property rights.

We refer to this situation as strategic delay. *Strategic delay* is the purposeful postponement of a violent or nonviolent settlement of a dispute with the hope of achieving a more preferable outcome in the future. In the bargaining literature, scholars have typically argued that strategic delay results from asymmetric information. When bargainers are uncertain of each other's preferences, they may have incentives to delay settlement in order to signal their bargaining strength.[35] In international property rights disputes, strategic delay is not necessarily the product of uncertainty about a disputant's relative strength. Instead, disputants may pursue delay because they anticipate that the strategic environment may change over time. Additionally, Wiegand finds that states may perpetuate these disputes because unsettled territorial claims can provide bargaining leverage in negotiations over other issues.[36] In general, strategic delay occurs when states are constrained from using violence but are unwilling to settle a property rights dispute. As Fravel argues, states often choose to delay because the other options are too costly.[37] In the context of our model of market power politics, states pursue a strategy of delay because they are not able to achieve a desired increase in market power either through violence or negotiation.

We identify three general situations in which states may choose strategic delay in the face of market power opportunities. First, a state that is not currently able to achieve a desired increase in market power through violence or negotiation may strategically delay in the hopes that its bargaining position will improve in the future. As we discussed earlier, states are generally not able to renegotiate boundary settlements. Thus, if a state were to agree to a dispute settlement that does not take advantage of a market power opportunity, it would likely preclude it from being able to achieve a more favorable settlement in the future. By delaying, the disputant is able to keep the market power opportunity on the table. If and when it finds itself in a more privileged position in the future—because a previous constraint to violence has been removed or because its bargaining power has increased—the state will be able to take advantage of the market power opportunity at that point.

---

35. Admati and Perry 1987; Cramton 1992.

36. Wiegand 2011.

37. Fravel 2008.

Second, strategic delay can provide opportunities for states to engage in gray zone tactics to pursue their market power goals. When states are constrained from pursuing violence, expansionist strategies short of war may still be on the table. For example, states can engage in salami tactics, which can allow them to slowly achieve increased market power over time. When using salami tactics, challenger states take small actions that give them greater access to territory or resources but do not sufficiently warrant an escalatory response by the target state. Target states that are highly dependent upon such a challenger may be unwilling to risk the costs involved with escalating the dispute. Through the continued use of salami tactics, states can take advantage of market power opportunities in two potential ways. First, they can slowly amass control over resources that would directly provide them with increased market power. Alternatively, states can use salami tactics to put themselves in a more favorable bargaining position, which would allow them to reach a dispute settlement that would provide them with a significant increase in market power.

Third, states that benefit from the status quo may engage in strategic delay to perpetuate uncertainty about property rights. In particular, states may delay to avoid a settlement that would give a rival state a significant increase in market power. Since property rights settlements are inherently difficult to revise, such an agreement would likely ensure that the rival state would be able to enjoy the benefits of this improved position in the market for an extended period of time. Instead, they may delay in hopes that a future shift in the strategic environment will either provide them with greater bargaining power or eliminate the rival's market power opportunity. In this way, delay can also be a tactic of market power prevention. Thus, by raising the stakes of any settlement, the institutional rules and procedures surrounding the delimitation of international property rights can potentially incentivize either a state facing a market power opportunity or its rival to pursue a strategy of strategic delay.

Figure 3.3 illustrates the conditions under which we expect to see an increased use of strategic delay in property rights disputes. When a high level of economic interdependence constrains a state motivated to pursue a market power opportunity, but the potential institutional settlements are not acceptable, that state will be more likely to strategically delay a resolution of the dispute. As in the cases of violent escalation, states facing market power opportunities have incentives to act aggressively to seize control of resources. However, a high level of economic interdependence between the disputants increases the cost of any armed conflict that they might fight. These anticipated costs constrain the disputants from using violence or even credibly threating to use military force. Economic constraints thus put the brakes on violent escalation. However, if international institutions are unable to effectively address or constrain the market power motivation, the

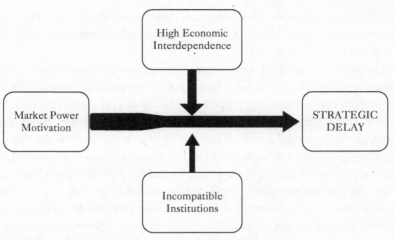

FIGURE 3.3. Conditions leading to strategic delay.

disputants may be unwilling to reach an acceptable settlement. This leaves the dispute in a holding pattern of strategic delay.

Dispute Settlement

While we expect that property rights disputes with market power opportunities will often be locked in a pattern of strategic delay and at times may escalate to violence, states will sometimes be able to peacefully resolve these disputes. As we have seen, international institutions play a critical role in facilitating successful settlements. By providing focal points and generating reputation costs, institutional rules and procedures can help states reach an agreement when there is a potentially mutually beneficial settlement on the table. Such a peaceful outcome will be more likely when additional factors, such as economic interdependence, constrain states from pursuing aggressive actions to achieve enhanced market power.

From our earlier discussion, we know that institutions can often be ineffective at dealing with property rights disputes when market power motivations come into play. Institutional rules can limit the set of potential settlements available to disputants. Additionally, since boundary settlements are often difficult to renegotiate, a challenger state desiring market power may be reluctant to sign an agreement that precludes a market power opportunity, while the target state may be unwilling to agree to a settlement that gives its rival long-lasting price-setting capabilities. Moreover, large shifts in market power can create commitment problems that can inhibit the ability of disputants to make use of the side payments or issue linkages needed to reach settlements. These factors shrink—and potentially eliminate—the bargaining range of mutually acceptable settlements available to disputants.

Despite these obstacles, international institutions can sometimes facilitate dispute settlement in the face of a market power opportunity. As we discussed in the previous chapter, resolving property rights disputes can create economic benefits by reducing transaction costs. In some cases, these economic benefits may outweigh the expected benefits of attempting to seize a market power opportunity through violence or strategic delay. For example, when economic interdependence raises the cost of pursuing violence and the probability of a future shift in the strategic environment is small, neither escalation nor delay may be viable strategies for a state to achieve a dominant market position. In these situations, states may be willing to accept an efficient resolution to a property rights dispute even if it precludes the possibility of a shift in market power. Additionally, since side payments and issue linkage will be more credible when a settlement does not create a market power shift, states may be better able to reach mutually acceptable agreements in these situations. Thus, we expect that most settlements that do arise will involve states giving up market power opportunities.

Figure 3.4 illustrates the conditions under which we expect that states will most likely be able to successfully settle property rights disputes with market power opportunities. As in the case of strategic delay, a high level of economic interdependence can constrain states with market power motivations from escalating to violence. When international institutions are sufficiently strong and can facilitate a mutually beneficial settlement, they can further dampen these market

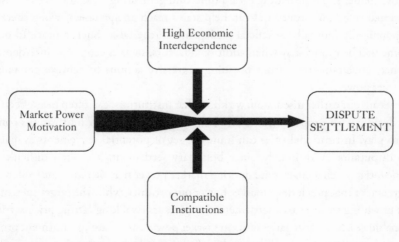

FIGURE 3.4. Conditions leading to dispute settlement.

power motivations. Disputants may then be willing to abandon a strategy of delay and instead agree to a peaceful dispute settlement. In such cases, the economic benefits of an efficient property rights settlement outweigh the potential gains that could be achieved from pursuing a market power opportunity.

As we have noted, institutions will be best able to facilitate peaceful dispute settlements that do not create a significant shift in market power since states face fewer commitment problems in these situations. This does not imply that it would be impossible for disputants to reach a settlement that gives one side a significant increase in market power. However, the conditions under which this could occur are likely to be narrow. In particular, disputants will likely only agree to a division of property rights that gives one state significant market power when institutional rules are in line with such a settlement and the challenger state has sufficient leverage to coerce the target into accepting such a settlement. Since this may require a credible military threat, such a settlement may be possible, and perhaps even more likely, without the constraint of economic interdependence.

Summary

Opportunities to set prices and shape the market can alter the behavior of states embroiled in property rights disputes. These market power opportunities motivate states to pursue aggressive strategies to gain control over the supply of valuable resources. How states respond to these motivations depends upon the level of constraints provided by economic interdependence and international institutions. High levels of economic interdependence increase the cost of economic exit and can constrain states from escalating to violence. In many situations, international institutions are not properly designed to handle issues that arise in the face of market power opportunities. However, when they are able to do so, institutional rules and procedures can facilitate the process of peaceful dispute resolution.

Table 3.1 summarizes our expectations as to how these economic and institutional constraints influence the outcomes of property rights disputes in the

Table 3.1. Theoretical Expectations about Dispute Outcomes

	Low Economic Interdependence	High Economic Interdependence
Incompatible Institutions	War	Strategic Delay
Compatible Institutions	Settlement or War	Settlement

face of market power opportunities. When there is a low level of economic inter-dependence and institutions are not designed to handle the issues surrounding market power, disputants are less constrained and are more likely to use violence to seize a market power opportunity. On the other hand, a high level of interde-pendence combined with incompatible institutions will provide some constraint on aggressive behavior and create incentives for disputants to pursue a strategy of delay. If interdependence is high and institutions are able to provide a mutu-ally beneficial agreement that outweighs the market power opportunity, peaceful dispute settlement is possible. Finally, our model does not provide a clear expec-tation about the outcome of disputes with low economic interdependence and compatible institutions. While the lack of economic constraints may allow for the possibility of violent escalation, the presence of compatible institutions may help disputants reach an efficient settlement in lieu of paying the costs of war.

Stepping back to view the bigger picture, the standard view of how insti-tutions matter in property rights dispute resolution is still intact; it accurately describes the causal path that occurs when market power opportunities are not associated with the territorial dispute. Two other important paths also come into focus, however. When market power opportunities motivate states to be more aggressive and there are no economic constraints in place, we should see an increased risk of armed conflict. When market power opportunities exist and economic interdependence is also present, we are more likely to see states use a strategy of delay. These outcomes of war and strategic delay are especially likely when potential institutional solutions are incompatible with the interests of states facing potential market power opportunities or their rivals.

Having outlined the expectations of our theoretical model of market power politics, we now turn to examining how these processes play out empirically. In Part II of the book, we explore a series of cases in which states have pursued expansionist tactics in pursuit of price-setting power in commodity markets tied to territory: Iraq's attempt to create a shift in the oil market through an invasion of Kuwait, Russia's ongoing efforts to maintain and expand its dominance of the regional natural gas market, and China's moves to control seabed resources in the South and East China Seas. In each of each of these cases, we show how the con-figuration of market power motivations, economic interdependence, and institu-tional constraints have influenced the decisions of these countries and their rivals to pursue strategies of violence or delay. In doing so, we illustrate how our model of market power politics can help us better understand the factors underlying some of the most high-profile international territorial and maritime disputes in the past few decades.

PART II

*Market Power Politics
in Commodity Markets*

4

Empirical Cases

UP TO THIS point, we have focused our attention on developing a theory to explain how international economic competition influences the expansionist activities of states. According to our model of market power politics, when a state's firms can set prices in key markets, the state can benefit both politically and economically. In addition to the potential for increased revenue from the rents gained by price-setting firms, market power can provide states with international bargaining leverage and stability. The desire to achieve such market power can motivate states to gain control of valuable territorial and maritime resources, potentially through the use of force.

However, these territorial ambitions do not necessarily lead to war, as economic interdependence can make military escalation too costly. Ideally, international institutions would help states efficiently resolve property rights disputes, but sometimes the rules outlined by these institutions preclude a settlement that is in line with a state's market power goals. In such cases, expansionist states may prefer to strategically delay any resolution of the dispute. By delaying, a state can either wait until it is in a better bargaining position before reaching a settlement, or it can try to gradually accumulate market power over time through salami tactics.

In the second part of the book, we turn our attention to exploring how these processes play out empirically. The three chapters that follow examine how market power motivations, in combination with varying levels of economic and institutional constraints, have shaped the expansionist strategies of three countries: Iraq, Russia, and China. In the case of Iraq, we examine how motivations to obtain price-setting capabilities in the oil market drove Iraq to invade Kuwait in 1990 and then turn its sights to Saudi Arabia, sparking the intervention of a US-led military coalition. We then discuss Russia's continuing efforts to maintain and increase its power in the European natural gas market, including its expansionist activities in neighboring Georgia and Ukraine. Finally, we consider how competition over seabed resources, including rare earth elements (REEs), shapes the patterns of strategic delay in the ongoing maritime disputes in the East and South China Seas.

Our goal in these empirical chapters is to demonstrate the plausibility of our theory of market power politics by examining some high-profile territorial and maritime disputes from the past few decades. In doing so, we aim to show how the theorized factors of market power motivation, economic interdependence, and international institutions have influenced the behavior of expansionist states and their rivals in these cases. In this analysis, we do not set out to provide a definitive test of our theory. Instead, we use these cases to illustrate how the causal mechanisms outlined in our theory play out in the real world. In this way, these case studies can be seen as plausibility probes. A plausibility probe is an empirical analysis undertaken after a set of hypotheses has been developed but before a full-fledged test is conducted.[1]

As probability probes, these cases help our understanding of market power politics in several ways. First, they provide an existence proof to demonstrate that there are empirical cases of property rights disputes where market power motivations have played a key role, as we have theorized. Second, the cases provide something akin to a pilot study that allows us to see if there is some empirical evidence that supports our hypotheses. In this way, these cases can be seen as the first step in an empirical analysis of market power politics in property rights disputes to determine whether further analysis is warranted. Third, they help us hone our theoretical argument and identify its scope conditions, as scientific knowledge is often best accumulated through an iterative process of theoretical development and empirical observation.

Our empirical discussion mainly consists of process tracing. We turn to such qualitative rather than quantitative analysis for several reasons. First, given that our goal is to identify the motivation behind the strategic decision-making in property rights disputes, qualitative case studies allow us to better empirically trace the causal mechanisms underlying this behavior. This is especially true when analyzing a complex, conditional theory like the one we have developed here. Additionally, since we do not argue that market power motivations drive all property rights disputes, it would be difficult to determine a priori a large sample of cases that would be relevant for our analysis. Finally, some of the key factors that we identify are difficult to systematically measure across cases. In particular, many of these factors are context-specific to the particular market in which a state

---

1. Eckstein 1975; Levy 2008.

is aiming to achieve price-setting capabilities. We now turn to discussing the criteria we used to select the cases that we analyze.

Case Selection

Given our desire to illustrate how states respond to market power opportunities, it would not be useful to randomly select cases to analyze. We do not anticipate that most states will have opportunities to secure price-setting capabilities, nor do we claim that such market power motivations will drive all territorial disputes. Thus, a random selection process would likely produce a set of cases in which market power motivations do not matter. While this would not necessarily be inconsistent with our theory, it would not help us evaluate the plausibility of the causal mechanisms that we outlined in the previous chapters. Instead, our approach is to identify cases where there is a possibility that our hypothesized processes could play out. That way we can better see how market power motivations, in combination with varying economic and institutional constraints, influence the strategies of territorial expansion.

With this in mind, we used the following three criteria to help select the empirical cases that we explore in the following chapters. First, we focus our attention on hard commodity markets in which a state could potentially have both the opportunity and willingness to pursue price-setting capabilities through territorial expansion. Second, to isolate the international market power mechanism, we simplify domestic political economy considerations by limiting our analysis to cases where the state has significant control over its firms in these markets. Finally, to observe the factors that influence strategic choice in the face of market power motivations, we select cases that provide variation on our dependent variable. Let us consider each of these criteria in turn.

Opportunity and Willingness: Hard Commodity Markets

Our empirical cases focus on markets for oil, gas, and REEs. In Chapter 3, we identified two conditions that are generally required for market power opportunities to motivate states to turn to violence or strategic delay. For one, these opportunities or threats should involve a key good for the state's economy. Additionally, shocks to the prices of these key goods must be costly in terms of adaptation by other players in the market, as this is the source of the political power of a market power opportunity. By focusing on these hard commodity markets, we are able to identify cases in which these conditions are likely to be met.

While market power competition is not limited to the trade of any particular good, the market power motivations that we identify only emerge when there is the possibility for an actor to obtain the ability to set prices. Additionally, these motivations will only influence strategic behavior in property rights disputes when changes in territorial and maritime control could lead to shifts in market power. Thus, to explore these processes empirically, we need to identify markets in which a state could have both the opportunity and willingness to pursue price-setting capabilities through territorial expansion. Markets for oil, gas, and REEs meet these criteria.

First, the structure of these commodity markets can provide actors with opportunities to gain price-setting capabilities. For each of these commodities, there are a small number of producers. This makes deviations from competitive markets more likely, as it is easier for an individual producer to gain control of a significant share of the available resources to set prices. Moreover, the geographic distribution of natural resources limits the entry of actors into these markets. To put it simply, most countries do not have access to significant amounts of these resources, and they cannot become producers unless they locate or gain control of new reserves. Additionally, production of these commodities requires significant investment in infrastructure for both extraction and transportation. On top of this, in some natural gas markets, it has been historically difficult for consumers to switch to new producers due to the fixed location of gas pipelines and the reliance on long-term contracts. All of these factors provide opportunities for producers to accumulate market power.

In addition, states can have the opportunity to increase their market power in these commodity markets through territorial expansion. Territorial and maritime control affects the ability to obtain and transport oil, gas, and REEs. If a state is able to gain control of territory or a maritime area that contains natural resources, it can increase its capability to produce them. At the same time, when a state gains access to such resources, it denies other states access to those resources. Thus, territorial expansion can potentially allow a state to increase its share of the market, which in turn can give it market power. Additionally, territorial expansion can potentially give a state greater control over the transportation of oil and gas, particularly when these resources are transported through pipelines. By vertically integrating the control of both the production and transit of resources, suppliers can increase their ability to set prices in a market. Additionally, such a move can inhibit the ability of other producers to transport resources to consumers. For these reasons, territorial expansion provides a mechanism by which producers can augment their power in these commodity markets.

Finally, international trade of oil, gas, and REEs is a highly salient political issue for both consumers and producers. For one, the supply of energy is a critical need for all states. Both oil and natural gas can be used to generate electricity and heat homes and businesses. Additionally, much of the world's transportation relies upon petroleum products, including gasoline, diesel fuel, jet fuel, and asphalt. Thus, successful economies require reliable and affordable access to energy sources, including oil and gas. At the same time, many hydrocarbon-producing states are highly dependent upon oil and gas exports, which often provide a significant share of government revenue. Rare earth elements are essential for the production of many commonly used consumer electronic devices, including cell phones, computers, and televisions. They also play an important role in many clean energy and defense industries. Thus, access to REEs is critical for any country engaged in advanced manufacturing in today's global economy. For these reasons, we expect that states would have significant incentives to fight over control of these critical hard-commodity markets.

State–Firm Relations

We also focus our empirical analysis on cases where the state plays an active role in its firms' activities in these markets. In doing so, we simplify the domestic relationship between firms and states. As we discussed in Chapter 3, we expect that, all else being equal, states that own or have significant control over their firms in a market will be more likely to take aggressive steps to gain or preserve market power than states with more liberalized economies. When a state owns or has significant control over its firms, there is less of a divergence between state and firm interests. Additionally, the state is in a better position to use its firms' price-setting capabilities as leverage to extract rents or achieve political goals. Thus, in these cases, we expect that there will be a more direct connection between market power motivations and foreign policy. By abstracting away from some domestic political economy interactions, we can better focus our attention on the international causal mechanisms outlined in our theory.

This is not to say that states and firms are completely aligned in the cases that we analyze. While an individual firm is primarily focused on its own success, state leaders have political goals on many dimensions. Thus, state leaders may sometimes prefer that a firm take steps that are in line with the government's political goals, even if it sacrifices the firm's profits. For example, to reduce its dependence upon Ukraine and Belarus as transit states, the Russian government has pushed Gazprom to pursue costly pipeline projects that significantly exceed its export capacity. While this may have long-term benefits, such efforts are financially risky and have at times led to short-term reductions in Gazprom's market value.[2] This could potentially lead to principal–agent problems in which a firm has incentives to take steps that deviate from the state's preferred strategy. However, these deviations will be less likely in these cases given the state's ability to exert greater influence over the firm's actions.

We recognize that these scope conditions for our empirical analysis limit our ability to extrapolate our findings to states with more liberalized economies. Even if we can show that market power motivations can drive the expansionist behavior of states that have a large degree of control over firms in a market, several open questions would remain as to whether similar dynamics would play out in more liberal domestic economies. Can firms motivate such states to pursue expansionist policies to increase the firms' market power? If so, how do these states reap the benefits from the firms' price-setting abilities? Does such market power provide states with the same political leverage in these cases? Answering these questions would likely require more theory development. Thus, for now, we will limit our attention to cases where states have a greater level of control over firm activities and will leave an empirical analysis of market power politics in liberalized states to future research.

Regardless of how far we can expand the scope of our analysis in the future, understanding the role of market power in the cases that we do examine is still important. For one, it provides us with a better theoretical understanding of how international economic relations can shape conflict behavior. Most of the literature on economics and conflict has focused on the role of trade flows between countries. By showing that market power motivations can drive expansionist foreign policies, we can demonstrate how other economic factors influence conflict behavior. Second, our case studies focus on some of the most important actors in the international system, including Russia and China. In the contemporary global environment, these are the major powers most actively engaged in expansionist territorial policies. They also pose the greatest potential challenge to the dominant role played by the United States in the post–Cold War era. Thus, it is important to understand their behavior.

Variation on the Dependent Variable

Finally, when selecting cases to analyze, we also want to make sure that there is variation on our dependent variable, the market power strategy that states pursue.

---

2. Vavilov 2015b, 9.

If we want to analyze the factors that influence strategy choice, we need to consider cases where states pursued different strategies. For example, if we only consider cases where states choose to go to war, we would not be able to see how the factors that lead states to pursue military escalation differ from those that lead them to opt for strategic delay. By ensuring that we consider cases in which states pursue different strategies in response to market power motivations, we are better able to see how variation in our hypothesized factors influences strategy choices in property rights disputes.

Our main focus in the case studies is to consider how states respond to market power opportunities. Given the constraints on the number of case studies that we can examine in this book, we have chosen to limit our attention to property rights disputes in which one state faces a market power opportunity and disputants pursue a strategy of war or delay. We choose to focus on war and strategic delay because these strategies represent the largest deviations from the expected behavior of states in normal economic exchange. Given this, our goal in the case studies is to first demonstrate the market power motivations faced by the disputants. We then discuss how our hypothesized factors of economic and institutional constraints influenced the choice to engage in either war or strategic delay.

While this approach allows us to illustrate the dynamics that influence the choice of violence and strategic delay in the shadow of market power opportunities, it limits our ability to evaluate other aspects of our theoretical argument. First, we do not consider cases where states successfully reached a peaceful dispute settlement in the presence of market power motivations. In the concluding chapter of the book, we do discuss a case of a successful negotiated settlement of a property rights dispute after a significant period of strategic delay (the Caspian Sea dispute) to illustrate how such agreements could come about. A more detailed examination of cases negotiated settlements would be a logical next step to examine the empirical implications of our theory.

Second, we do not compare cases with and without the presence of market power opportunities. In our theory, we do not claim that market power opportunities provide the only reason that property rights disputes emerge. Instead, we argue that market power ambition is an underexplored factor that motivates some expansionist behavior in international relations. For this reason, our primary goal in these cases is to demonstrate how this process plays out empirically. Thus, while a full test of our theory would require us to examine the effects of change in market power motivations, we have chosen to leave that to future research.

Overview of the Cases

In the next three chapters, we focus on the expansionist activities of three states aiming to amass market power in key commodity markets: Iraq, Russia, and China. For each of these states, we examine the strategies they took to try to gain or maintain price-setting capabilities. Table 4.1 outlines the specific cases of market power pursuit that we will explore in detail. For each case, we list the relevant commodity market, the targeted territory, and the primary strategy used by the state seeking market power. In line with the criteria we outlined in the preceding section, we include cases of both war and strategic delay to provide variation on the dependent variable.

In addition to these cases of market power pursuit, we also consider the strategies of rivals to prevent these three states from achieving their market power goals. For two of these countries, Iraq and China, we examine actions that rival states took to prevent them from expanding their territorial control over key resources. Table 4.2 outlines these cases of market power prevention, which we will investigate in Chapters 5 and 7. To provide variation on the dependent variable, they include a case of war and a case of strategic delay. While we do not

Table 4.1. Cases of Market Power Pursuit

State Seeking Market Power	Commodity	Territory/Maritime Area	Primary Strategy
Iraq	Oil	Kuwait	War
Russia	Natural Gas	Georgia (South Ossetia, Abkhazia); Ukraine (Crimea, Donetsk, Luhansk)	Strategic Delay
China	Rare Earth Elements	East and South China Seas	Strategic Delay

Table 4.2. Cases of Market Power Prevention

State Seeking Market Power	State Preventing Market Power	Territory/Maritime Area	Primary Strategy
Iraq	United States	Northeastern Saudi Arabia	War
China	Japan	East China Sea	Strategic Delay

consider territorial strategies of market power prevention in the case of Russia, we do discuss some of the various strategies taken by European countries to limit Russia's price-setting abilities in Chapter 6.

To demonstrate how these cases fit the selection criteria that we have outlined, we provide a brief overview of each case in the following subsections. First, we explain the importance of the relevant commodity for the state that is seeking market power. Then, we describe the nature of the relationship between the state and its firms in the commodity market. Finally, we provide a brief description of the market power opportunity and the strategies taken by the state and its rivals.

Iraq

In Chapter 5, we discuss Iraq's attempts to increase its market power in the oil export market in the early 1990s. At the time, Iraq was the second-largest oil exporter in the Organization of the Petroleum Exporting Countries (OPEC), producing almost 13 percent of the cartel's output.[3] As a petro-state, Saddam Hussein's Iraq largely depended upon the sale of crude oil, which made up the vast majority of its exports. Overall, this oil trade accounted for more than 60 percent of Iraqi gross domestic product and 95 percent of its foreign currency earnings.[4] Access to oil revenue was critically important for Hussein's regime given Iraq's significant debt in the wake of the costly war with Iran that ended in 1988.

Under Saddam Hussein, the state had direct control over the oil industry in Iraq. The Iraqi government had nationalized the oil industry in the early 1970s. In 1987, under a major restructuring of the oil sector, the Iraq National Oil Company (INOC) was merged with the Oil Ministry, which divided INOC's subsidiaries into separate companies.[5] At this point, all sectors of the Iraqi oil industry were placed directly under the jurisdiction of the oil minister. Thus, during the period we examine, Iraqi oil companies were state-owned enterprises under the direction of the Oil Ministry, and oil was the main revenue source for the state. Given this, gaining price-setting capabilities in the oil market would directly benefit Hussein's government.

We consider two opportunities for Iraq to increase its market power in the oil market through territorial expansion. First, in a case of market power pursuit, Iraq opted to use military force when it invaded its oil-rich neighbor, Kuwait. By controlling Kuwait's oil production, Iraq could both augment its own oil

3. Williams 1999.

4. Duelfer 2004, 207.

5. Jaffe 2006.

resources and prevent Kuwait from overproducing and putting downward pressure on the price of oil. The second market power opportunity would have been for Iraq to subsequently invade Saudi Arabia, giving Iraq a sufficient market share to be able to control the global output and price of oil. To prevent this shift in market power, a coalition of forces led by the United States intervened militarily to stop Iraq from occupying Saudi Arabia. This provides an example of the use of war as a strategy of market power prevention.

Russia

In Chapter 6, we examine Russia's strategies to maintain and potentially expand its privileged position in the European gas market. Natural gas plays a critical role in the Russian economy. Russia has the largest proven gas reserves in the world, and it supplies 18 percent of global gas production. The majority of Russia's export revenues come from energy products, and natural gas accounts for 12 percent of the country's total exports. The Russian state is highly dependent upon hydrocarbons, as over 40 percent of its federal government revenues come from oil and gas.[6]

Gazprom, the Russian gas company, is a vertically integrated firm that owns the world's largest gas reserves and the world's largest gas transmission network. In the 1960s, the Soviet Union's Ministry of Gas Industry was in control of natural gas exploration and extraction. Gazprom emerged from this ministry as the Soviet Union's first state-owned enterprise in 1989. When the Soviet Union dissolved, Gazprom went through a decade of privatization, but the Russian state maintained a significant stake in the publicly traded company. Under Putin, the Russian state gained a majority stake in Gazprom, and in 2006, the government gave Gazprom a legal monopoly over Russian gas exports. Through its control of Gazprom, the Russian state has been able to use the firm to achieve domestic and international political objectives and to extract rents from gas exports to provide a stream of government revenue.[7]

Due to its position as the primary supplier of natural gas to many European countries, Gazprom has significant market power in the regional gas market. Given the political and economic benefits of price-setting capabilities, Russia has a strong motivation to take steps to protect and potentially enhance this market power. In territorial disputes with its neighbors, Russia has mainly relied upon strategic delay to pursue its market power goals. For one, it has perpetuated

---

6. BP 2019; World Bank 2019.

7. Vavilov 2015a.

the frozen conflicts in South Ossetia and Abkhazia with the aim of destabilizing Georgia, which is a critical transit state for a rival gas producer (Azerbaijan). At the same time, Russia has largely opted to use gray zone tactics to expand its influence in the Crimea and eastern Ukraine, which increases Russia's leverage over its own transit state, Ukraine, and provides access to new gas reserves in the Black Sea.

China

In Chapter 7, we turn to our attention to China's efforts to maintain its ability to set prices in markets for REEs. Rare earth elements play a critical role in China's economy. First of all, China is the dominant global producer of REEs. For decades, China produced nearly all of the REEs in the global economy, and today it produces four times the amount of REEs as all other producers combined. However, not only is China the largest producer of REEs, it is also the largest world's largest consumer. Downstream industries require REEs for the production of many advanced electronics and "green energy" products. As China moves to update its manufacturing capabilities in these high-tech sectors as part of its Made in China 2025 blueprint, REEs will play a critical role in the government's economic goals of leaping over the middle-income trap and transitioning to a high-income economy.

The Chinese government plays a central role in its domestic REE production. Over time, Beijing has implemented policies to enhance its control over these industries. The state controls the total quantity of REE production, assigns production quotas to firms, and approves all rare earths projects.[8] While the REE industry in China was historically fragmented, in recent years the state has encouraged consolidation of the industry through mergers, particularly in the upstream and midstream sectors. By 2017, almost all mining and separation of REEs in China had been concentrated in six state-owned enterprises.[9] Given the domestic policy environment and state ownership of firms, the Chinese state is in a strong position both to control and to benefit from the country's production of REEs.

As the dominant producer of REEs in the world, China is a price setter in the market. The Chinese government has a strong motivation to maintain this position, as it allows China to be the first choice for producers using these minerals throughout the global value chain while maintaining stable and competitive

---

8. Shen et al. 2017.

9. Shen, Moomy, and Eggert 2020.

access to rare earths for China's domestic economy. As part of its efforts to preserve this market power, China has turned to strategic delay and gray zone tactics to continue to push its claims for sovereignty in the South and East China Seas. By expanding sovereign control over these seabed resources, China would have access to a steady supply of REEs to extract for the next few decades. At the same time, as a major producer of advanced electronics, Japan has used strategic delay tactics to prevent any settlement of the East China Sea dispute that would deepen China's monopolistic power in this hard commodity market.

Summary

In the second part of the book, we conduct a series of case studies to explore how the theoretical processes outlined in our model of market power politics play out empirically. In Chapter 5, we examine how the motivation to gain price-setting capabilities in the oil market led Iraq to invade Kuwait and set its sights on Saudi Arabia, which prompted a military intervention by a coalition of forces led by the United States. Chapter 6 explores the portfolio of strategies that Russia uses to maintain and enhance its market power in the regional gas market, including the use of various gray zone tactics in Georgia and Ukraine. Finally, in Chapter 7, we analyze how competition over seabed resources between China and its regional neighbors has driven the patterns of strategic delay in the long-standing maritime disputes in the East and South China Seas.

We selected these cases because they provide useful plausibility probes for our theory of market power politics. First, they involve competition in key commodity markets in which states could potentially have the opportunity and willingness to pursue a market power opportunity though territorial expansion. Second, since these states have had significant control over their firms in these commodity markets, we are able to isolate the mechanism by which market power motivations influence foreign policy decisions. Finally, these cases include incidents of both violence and strategic delay, which allows us to examine how variation in economic and institutional constraints influences strategy choices in property rights disputes.

Each of our empirical chapters follows a similar pattern. First, we provide a brief history of the case and describe the opportunity for a state's firms to obtain price-setting capabilities in a key commodity market. We then examine how this opportunity motivated the state to pursue expansionist strategies to enhance its market power and in turn motivated rivals to take steps to prevent it from doing so. With this market power motivation established, we then consider how the configuration of economic and institutional constraints influenced the choice to pursue strategies of violence or strategic delay. We conclude each chapter with a discussion of alternative causal mechanisms to see how well our theory of market power politics holds up to competing explanations for these countries' actions.

5

Iraq

FIGHTING FOR MARKET POWER

THE WORLD REACTED with surprise when Saddam Hussein gave the order to invade Kuwait in the middle of the night on August 2, 1990. Hussein had made promises not to invade, assuring that he would keep his troops at the border as negotiations continued between Iraqi leaders and delegations from Egypt, Saudi Arabia, and even the United States. Part of this surprise was undoubtedly due to the prior decade of war between Iraq and Iran. This war of attrition had cost Iraq over half a million casualties, hobbled its oil production, and was tremendously expensive.[1] Prior to the Iran-Iraq War in 1980, Iraq had over $35 billion in foreign reserves. By 1988, however, Iraq's foreign debts exceeded $100 billion.[2] Given the brutal costs of Iraq's previous war with Iran, few expected Saddam Hussein to have the appetite for more war. Yet Iraq launched an invasion into Kuwait using over 100,000 troops. Figure 5.1 illustrates the invasion path taken by Iraqi forces. The operation was quick and effective, and within a day Iraqi forces controlled the entire state of Kuwait.

Iraq's success, however, was fleeting. Within a week of the invasion, the United States began sending forces to Saudi Arabia. When economic sanctions and diplomatic efforts failed to pressure Iraq to leave Kuwait in the ensuing months, a coalition of forces led by the United States launched an air assault on Iraqi forces on January 17, 1991. A ground offensive ensued on February 24, and the Coalition forces successfully pushed the Iraqis out of Kuwait within 100 hours. By the end of February 1991, the Persian Gulf War was over. The Iraqi army was soundly defeated, and Hussein's expansionist plans went up in flames. Given this eventual

---

1. Dawisha 2009.

2. Ibid., 223.

FIGURE 5.1. Iraqi forces invading Kuwait.

Copyright: Ronald J. Brown, *U.S. Marines in the Persian Gulf, 1990–1991: With Marine Forces Afloat in Desert Shield and Desert Storm*, U.S. Marine Corps, History and Museums Division, Washington, DC, 1998.

failure, one might wonder what led Saddam Hussein to take such a costly gamble of invading his southern neighbor in the first place.

In this chapter, we examine how the motivation to establish market power in the oil export market, in combination with few economic or institutional constraints, influenced Saddam Hussein's surprising decision to attack and occupy Kuwait. At the time, Iraq desperately needed new resources that could not be generated without obtaining price-setting capabilities. Because oil prices fluctuate with the typical dynamics of a commodities market, Iraq could not simply increase its own oil production to achieve its needed revenue targets. For years prior to the invasion, Iraq lobbied fellow partners in the Organization of the Petroleum Exporting Countries (OPEC) to reduce their production in order to raise prices. These lobbying efforts largely failed, and Kuwait and the United Arab Emirates (UAE) routinely exceeded their targets. With the market saturated with OPEC oil, an increase in production by Iraq would have further depressed oil prices, offsetting much of the revenue gained from increased output.[3] Seizing Kuwait's significant oil resources provided an alternative way for Iraq to improve its position in the oil market and potentially be able to affect oil prices.

As this market power opportunity motivated Iraq to expand its territorial reach over additional oil resources, Iraq faced few economic or institutional constraints on the use of force. Economic interdependence was insufficiently present to tie Iraq's hands. Outside of their cartel cooperation in oil exports, Kuwait and Iraq traded very little. Their exports and imports were largely similar, making them competitors without interconnected economies outside of oil. Moreover, Iraq's diversity of import resources meant no single country could leverage its economic ties to deter Iraq from invading Kuwait.

At the same time, international institutions failed to provide non-militarized solutions for Iraq's resource needs. The cartel-style regime of OPEC failed to provide Iraq the mechanism it sought to rein in oil output by Saudi Arabia, Kuwait, and the United Arab Emirates. Additionally, Iraq's legal claim over Kuwait was weak, and the design of organizations like the Arab League and the United Nations (UN) failed to provide procedures to effectively prevent the conflict. Given the absence of significant economic and institutional constraints that might have put brakes on the process, Iraq's market power motivations pushed Saddam Hussein to decide to invade Kuwait.

In addition to our analysis of the Iraqi invasion, we also explain how the decision of the United States and its Coalition partners to repel Iraq from Kuwait provides an example of the use of force to prevent a significant shift in market power. While the initial foray into Kuwait would have generated a significant increase in Iraq's market power, it paled in comparison to the leap in price-setting power Hussein would have enjoyed were he able to control or disrupt Saudi oil production. After the clear demonstration of false promises that preceded the first invasion, Saudi and American leaders feared a second invasion was imminent. This convinced the United States that Saddam Hussein had to be stopped through military means rather than through the use of international institutions or economic sanctions. In response, the United States organized a coalition of forces to push Iraq out of Kuwait and prevent Iraq from establishing near-hegemonic market power in the production of oil.

---

3. Mabro 1992.

As with the other cases explored in this book, we do not argue that market power was the only cause for the political outcomes that we observed in Iraq and Kuwait. Our analysis is akin to a partial-equilibrium approach, highlighting the exacerbating role of the market power opportunities that made a negotiated settlement much more difficult to achieve. Such a disclaimer is especially relevant here when examining the use of force. The decision to go to war is complex, and our interest lies in demonstrating how market power politics contributed to that decision, rather than claiming a complete explanation of Saddam Hussein's strategy choices. Nevertheless, we do consider several alternative explanations for Iraqi behavior and show that they are either less compelling than or consistent with one based upon market power.

With this caveat in place, we proceed with a brief overview of the history and tensions between Iraq and Kuwait. We then identify the presence of a market power opportunity for Iraq that served as motivation to horizontally integrate Kuwait into Iraq's own oil production capacity, as well as the second market power opportunity to absorb some of Saudi Arabia's oil capacity. After considering these market power motivations, we show that Iraq faced few economic or institutional constraints on using force. This imbalance between motive and constraint helps to explain the decisions of both Iraq and the US-led Coalition to use force. Lastly, we examine some alternative causal mechanisms that may compete with our explanation of Iraq's invasion of Kuwait and the subsequent Persian Gulf War.

Iraq and Kuwait: Historical Perspective

One could characterize the pre-oil relations between Iraq and Kuwait as always close by, but never close. The two states shared a history of being co-governed, first under the Ottoman Empire throughout the eighteenth and nineteenth centuries, then under British influence. British forces governed the region through the transition from Turkish rule and adjudicated local disputes over territory after the lifting of Ottoman rule. In November 1922, Sir Percy Cox, British High Commissioner to Iraq, convened a meeting of Iraqi, Saudi, and Kuwaiti leaders at the fort of Uqair to establish formal boundaries for these three states.[4] Cox drew up the borders on a map during the meeting itself, including the creation of neutral zones to serve as a buffer

---

4. Casey 2007, 55–56.

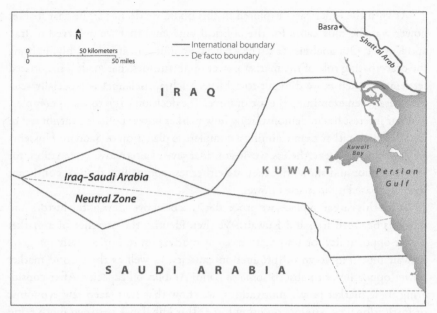

FIGURE 5.2. Neutral zone between Iraq and Saudi Arabia.

between Saudi Arabia and Iraq, as well as between Saudi Arabia and Kuwait (see Figure 5.2).[5]

Like much of the region at that time, Iraq emerged as a state in the post–World War I era under the close supervision of the British.[6] The British assumed control of the territory from the Ottoman Empire in the spring of 1917 and initially had little inclination to craft an Iraqi state. Internal British debate simmered over Woodrow Wilson's push for self-determination around the world, and in 1920, British opinion shifted to openness to establishing a constitutional monarchy. It would take over a decade for Iraq to take full advantage of this shift and form statehood in 1932. Part of the delay was due to tensions among Shi'a, Kurdish, and Sunni communities, which extended back to this first government; as the concept of coexistence within a centralized government proved to be a bigger challenge than the British expected. The nascent state limped through

---

5. The notion of these neutral zones emerges from controversy over Article VII in the 1913 Anglo-Turkish agreement that recognized the special relationship between Great Britain and Kuwait and laid out initial boundaries for Kuwait. See El Ghoneimy 1966, 691–692. Of course, Saudi Arabia is not yet a state at this time either, and is referred to in these documents as the Kingdom of Najd.

6. This historical summary is culled from Dawisha 2009. See especially Chapters 1 and 10.

wobbly regimes for decades, until a dual-phased military coup in the summer of 1968 finally installed a Ba'athist regime that consolidated power around Ahmad Hassan al-Bakr, aided by his close second-in-command, the young Saddam Hussein.

Saddam Hussein quickly established his position as an ambitious and ruthless political force. Dawisha characterizes the young leader dramatically, writing that as an "adroit political manipulator and determined street fighter, Saddam was never concerned about getting his hands dirty with the blood of those he considered a danger to the Party generally, but mainly to himself."[7] In 1979, Hussein assumed the presidency, succeeding President Bakr following his resignation. Hussein's first act was to root out top leaders who were sympathetic to a supposed plan to unify with their western rival, Syria, and the rival Syrian Ba'ath party.[8] A little more than a year later, Hussein launched Iraqi forces into the southwestern corner of Iran, hoping to control the Shatt al-Arab river that both states used to transport oil to the Persian Gulf. An expectation of swift victory proved to be misguided, and the ensuing war of attrition became another infamous proxy battleground in the Cold War between the United States and the Soviet Union. The brutal war would bring both economies to their knees, ending in a stalemate in 1988. Upon realizing that he would never control the oil resources of Iran, Hussein turned his attention to Kuwait.

Kuwait itself came into existence in the early eighteenth century, settled by the Bani Utub, and shortly thereafter emerged as a small and unassuming vassal state of the Ottoman Empire.[9] As in Iraq, the British assumed control of Kuwait following their victory over the Ottoman Empire in the First World War. Economically, Kuwait struggled in the years after World War I. Before oil transformed the little state, Kuwait relied principally on its pearl-harvesting industry, along with small amounts of fishing and agriculture. A glut in the global supply of pearls in the 1930s only magnified the stark effects of the Great Depression. Oil was discovered in the 1930s but would not be a steady source of income for the powerful families of Kuwait until the end of World War II. Soon thereafter, Kuwaitis would discover a seemingly limitless supply of oil, with revenues to match. As Casey writes, the "ability of Kuwait's sheikhs to spend most of those funds wisely would shape Kuwait for the rest of the century."[10] Income translated

---

7. Dawisha 2009, 211.

8. Ibid., 213–214.

9. This historical summary of Kuwait is culled from Casey 2007.

10. Casey 2007, 59.

to power and influence, ultimately transforming this sleepy pearl-harvesting corner of the world into an economic powerhouse.

The two states of Iraq and Kuwait also have a common history in terms of
how oil transformed their economies, and once oil was discovered in Kuwait,
Iraq's interest in its small neighbor increased markedly. Kuwait formalized its
independence in 1961 and joined the Arab League. The Iraqi regime protested
when Kuwait petitioned the British for independence, claiming that Kuwait
should not be granted independence but should instead be a part of the southern
Iraqi province of Basra. Kuwait's ties with Basra date to its days in the Ottoman
Empire, but only in that Basra was the local source of Ottoman authority and
contact for the Kuwaiti settlement. These protests continued until Iraq became
distracted by its war with Iran in 1979. Kuwait allied itself with Iraq and provided supplies and loans throughout the war. While Iraq never formally relin-
quished its desire to expand its borders to the southeast, the alliance between
the two states during the Iran-Iraq war seemed to indicate a normalization of
relations.

That short-lived normalization would not last, however. In the next section,
we detail the motivation behind Iraq's focus on Kuwait following the Iran-Iraq
War. The combination of massive debt and few options to generate much-needed
revenue, outside of profits from oil, would dominate Saddam Hussein's calculations. When war with Iran ended in stalemate, Hussein needed a new path to
shore up his economy and his own survival in the Iraqi government.

Motivation: Price-Setting in the Oil Market and Iraqi Debt in 1990

As Iraq limped away from a devastating decade of conflict with Iran, it was forced
to attend to its damaged economy. Oil exports during the war were diminished
due in part to conflict over oil transportation routes through the Shatt al-Arab,
the river border that both states used to move oil into the Persian Gulf. Iran had
destroyed Iraq's main oil terminals and refineries in the south early in the war,
and "daily oil production had fallen ominously from 3.4 million barrels in August
1980 to 800 thousand barrels exactly a year later."[11] Dawisha estimates that oil
production remained at roughly a third of pre-war levels throughout the war
with Iran. In the aftermath of the end of the war, reconstruction costs for Iraq
were estimated at $230 billion.[12]

---

11. Dawisha 2009, 223.

12. Karsh and Rautsi 1991, 19.

Ultimately, the basic motivation to invade Kuwait was Iraq's domestic political and economic environment. Gause notes that, rather than being a diversionary war to increase Hussein's domestic popularity, the invasion was aimed at stemming what Hussein saw as international efforts to weaken him domestically.[13] From Hussein's perspective, his fundamental domestic problem was his own political instability, which was exacerbated by the economic costs of the war with Iran but was rooted in regional influences that instantiated in Iraq as tensions between Shi'a, Kurds, and the ruling Ba'ath party. Without healthy oil revenue streams and hobbled by debt from the war, Hussein felt ill-equipped to fight off domestic threats, and the issue became one of political survival.[14]

This basic motivation translates into Hussein's evaluation of two distinct market power opportunities that we detail in the following subsections. First, by invading Kuwait, Hussein could resolve Kuwait's intransigence with respect to OPEC quotas by horizontally integrating his competition to the south. The second market power opportunity would have been to subsequently invade Saudi Arabia in order to capture a sufficient market share to control the global output and price of oil. Fear of such an outcome caused major powers in the West to mobilize and ultimately thwart Hussein's plans. When Hussein's attempts to seize these market power opportunities failed, he then turned to a last-ditch effort to improve Iraq's competitiveness by dismantling Kuwait's oil infrastructure as his army retreated.

Iraq's Invasion of Kuwait: A Quest to Control the Market

With oil revenues decreased throughout the conflict with Iran, Saddam Hussein had turned to foreign loans to finance the war. Despite Kuwait's support as an ally throughout the Iran-Iraq war, Iraq's focus shifted to Kuwait as a solution to its economic woes. Already deep in debt from the war, adding new debt without an established source of revenue would be difficult. In the summer of 1990, Iraq resurrected its old challenges concerning its border with Kuwait. Iraq argued that the state of Kuwait was never truly independent and was always properly considered a part of the Basra province of Iraq. Furthermore, Iraq asserted that even if Kuwait was independent, the borders between Iraq and Kuwait had never been properly established.

Iraq also protested Kuwait's overproduction of oil (exceeding the current OPEC quota) and demanded that Kuwait forgive Iraq's $13 billion war debt.

---

13. Gause 2002, 49.

14. Bueno de Mesquita and Siverson 1995; Gause 2002; Karsh and Rautsi 1991; Dawisha 2009.

This debt forgiveness was based on the notion that Saddam was fighting a war on behalf of the Arab world against fundamentalist Islam. Karsh and Rautsi note that during the Iran-Iraq War, Saddam pressured the Gulf States, Saudi Arabia, and Kuwait for loan forgiveness because he did not believe that they should be able to free ride on Iraq's actions.[15] These claims grew bolder after the war.

In addition to debt relief, Hussein set out to curtail the world's oil production in order to increase the global price. Both Iran and Iraq made demands upon the rest of OPEC to curtail their oil production so that these two recovering states could increase their own production without depressing prices.[16] Hussein then hosted an Arab summit meeting in May 1990, where he made his pitch explicitly:

> "For every single dollar drop in the price of a barrel of oil," he told his guests, "our loss amounts to $1 billion a year. Is the Arab nation in a position to endure a loss of tens of billions as a result of unjustified mistakes by some technicians or non-technicians, especially as the oil markets, or let us say the clients are, at least, prepared to pay up to $25 for the next two years, as we have learned or heard from the Westerners who are the main clients in the oil market?"[17]

Iraq's economic recovery would take decades when held back by the low oil prices of the first half of 1990. Most importantly, Iraq claimed that Kuwait was illegally slant drilling at its "border," effectively stealing from Iraq's reserves to the tune of $2.4 billion in lost revenue.[18] Iraq's foreign minister, Tariq Aziz, accused the Kuwaitis of "setting up oil installations in the southern section of the Iraqi al-Rumaila oil field and extracting oil from it." Moreover, Aziz accused Kuwait and the UAE of having "implemented an intentional scheme to glut the oil market with a quantity of oil that exceeded their quotas as fixed by OPEC." Such activities had allegedly resulted in a decade-long theft of resources, and Aziz went so far as to argue that "the drop in oil prices between 1981 and 1990 led to a loss of $500 billion by the Arab states, of which Iraq sustained $89 billion."[19] Given Iraq's position on the front lines against Iran, Aziz argued that the rest of the Arab world should compensate Iraq accordingly.

---

15. Karsh and Rautsi 1991, 21.

16. Ibid.

17. Ibid., 22. We assume the $25 refers to price per barrel.

18. Casey 2007; Crystal 1995; Finnie 1992.

19. Quotes sourced from Karsh and Rautsi 1991, 25–26.

There is indeed evidence that Kuwait's oil activities were troublesome for Iraq and other OPEC members, though the slant-drilling allegation was impossible to substantiate. Instead, it is uncontested that both Kuwait and the UAE were consistently exceeding their OPEC quotas. From Iraq's perspective, they were flooding the market. This overproduction by Kuwait and the UAE led to a steep reduction in oil prices in the six months of 1990. From January to June, oil prices fell from three-year highs to the lowest prices since 1986. Prices for West Texas Intermediate, the benchmark crude, sold for as high as $23.40/bbl and as low as $15.30/bbl on the spot market.[20] The decision by Kuwait and the UAE to exceed their OPEC quotas angered Iraqi leaders. As Verleger notes, "By the end of June, President Hussein and his oil minister Issam Abdul-Rahim al-Chalaby were making their feelings public. The price of oil began to increase in spite of the oversupply."[21] The increases were minimal, however, and did little to assuage Iraq's worries that it would be able to use its own oil revenues to service its debts.

Negotiations to resolve the dispute between Iraq and Kuwait were in place for months before the invasion. Beginning in May, after Iraq's initial demands against Kuwait, a set of parallel bilateral negotiations took place between Iraq and Kuwait, Saudi Arabia, Egypt, and the United States. Kuwait initially responded to Iraq's demands with incredulity, refusing to make any concessions. Attempts by Egypt, Saudi Arabia, and the United States to mitigate Hussein's concerns, including offering pecuniary rewards to Iraq if it backed down from the crisis, also failed to resolve the dispute. Deliberations in July intensified as Iraqi forces amassed on the border with Kuwait. As more and more forces arrived, Iraq's demands and tenor seemed to become increasingly aggressive. On July 17, Iraq threatened to use force against Kuwait and the UAE if they did not reduce their oil production. As tensions rose, Egypt's Hosni Mubarak and Jordan's King Hussein both attempted to mediate the dispute.

In response to the aggressive posture of the Iraqi regime, Kuwait began to yield to some of Iraq's demands. On July 25, the Kuwaitis agreed to price targets of at least $20/barrel and to reduce their output from over 2 million barrels per day to a target of 1.5 million.[22] On July 31, Kuwait agreed to forgive the war debt permanently and offered an additional $9 billion in loans to Iraq, and the Saudis offered financial support as well.[23] Other reports suggest that Kuwait sought a

---

20. Verleger 1990, 1.

21. Ibid.

22. Ibrahim 1990.

23. Whether Kuwait actually offered debt forgiveness and new loans is a matter of debate. Casey (2007, 87–88) states that on July 31, Kuwait agreed to the forgiveness of $13 billion in debts,

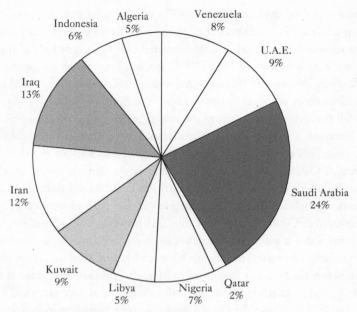

FIGURE 5.3. OPEC production, January 1990.
Source: Williams 1999.

compromise payment on Iraq's claim that Kuwait had stolen $2.4 billion in oil through slant-drilling. There is no evidence, however, that Kuwait agreed to border changes, especially with respect to islands in the Gulf.

While the threat of the use of force may have led Kuwait (and the UAE) to agree to lower their oil output, this only provided a short-term solution to the issue of overproduction. The inherent commitment problem would remain, and Kuwait would continue to have incentives to exceed its OPEC quota in the future. An alternative way to deal with Kuwait's intransigence with respect to OPEC quotas would be to horizontally integrate the competition.[24] By annexing Kuwait, Iraq would gain control of additional oil reserves that would generate a significant increase in market power. As Figure 5.3 shows, if Iraq could combine Kuwait's oil production with its own, it could nearly rival the output of the Saudis. In doing so, Iraq had the potential to shift away from its current economic environment—in which it had to take the price of oil as negotiated by the cartel

---

plus an additional $9 billion in interest-free loans. Others write that Kuwait never agreed to forgive old debts and only agreed to $1.5 billion in payments to settle the slant-drilling dispute.

24. Williamson 1975, 1985, 1996.

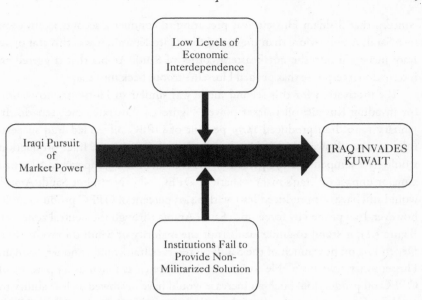

FIGURE 5.4. Iraq's motive and insufficient constraint.

and backed by the swing producing state, Saudi Arabia—and into a price-setting environment that would give Hussein bargaining power within the cartel.

This market power opportunity helps to explain why Iraq pushed forward with military action despite Kuwait's concessions (see Figure 5.4). While Kuwait agreed to reduce its oil production in line with its OPEC quota, it was not willing to unilaterally yield territory that would hobble its own access to the Persian Gulf while providing new access to Iraq. Additionally, the pecuniary rewards offered by countries like Egypt, Saudi Arabia, and the United States if Iraq backed down from the crisis were insufficient to deter Iraq from invading Kuwait. Iraq expected to receive all of these rewards upon successfully invading Kuwait. Moreover, the military imbalance in favor of Iraq was immense. Negotiations between Iraq and Kuwait broke down for the final time on August 1, and Iraq proceeded to take military action against its southern neighbor. The invasion began between midnight and 1:00 a.m. on August 2. Hussein's forces controlled Kuwait by mid-day.

Operation Desert Storm: Thwarting Iraq's Second Power Grab

Iraq's invasion of Kuwait surprised observers around the globe and drew attention to the seriousness of Saddam Hussein's aggressive posture. Of chief concern to local states and to major powers such as the United States was the fear that Iraqi forces would not stop with Kuwait. Specifically, there was widespread concern that Saddam Hussein was preparing to conduct a secondary invasion into Saudi Arabia. More than the initial foray into Kuwait, it was this fear of an Iraqi incursion into the northeastern corner of Saudi Arabia that triggered an international response that pushed Hussein's troops back into Iraq.

The motivation for this second move was similar to Hussein's motivation for invading Kuwait: oil market power. Figure 5.3 illustrates the scenario. In January 1990, Iraq produced 12.79 percent of OPEC oil.[25] Had Iraq success-fully annexed Kuwait and retained normal production rates, Iraqi production would have jumped to 21.49 percent of OPEC oil. Such a jump would have had a major impact on Iraq's market share of OPEC oil exports, but Saudi Arabia would still have out-produced Iraq, with 24.03 percent of OPEC production. If, however, Iraq pushed its forces into Saudi Arabia through the neutral zone (see Figure 5.2), it stood to annex or disrupt the majority of Saudi oil production. Depending on how much of the Saudi oil reserves Iraq could conquer, Saddam Hussein may have been able to increase his share to as much as 45 percent of OPEC oil production. Such an increase would have bestowed a clear ability to set the price of oil by controlling the global supply of oil coming out of OPEC states. The combination of the cartel nature of OPEC and Iraq's new strength within the group would have put Hussein in a position to manipulate global oil supply levels.

The threat of invasion was all too real for Saudi leadership. On the day of Iraq's invasion of Kuwait, Saudi leaders sent Prince Bandar bin Sultan bin Abdulaziz to meet directly with US President George H. W. Bush.[26] Bandar's mission was to ascertain whether the United States was willing to formally and fully commit military assistance to Saudi Arabia. Given the domestic political costs of involving the United States, King Fahd was reluctant to make a formal request unless he knew it would be accepted. Bandar traveled to the White House to meet with President Bush personally to ascertain the Americans' intent.

Bandar held important information about the morning activities of the Iraqi military. In particular, Iraqi tanks had entered the neutral zone between Iraq and Saudi Arabia three times, following what was clearly a natural route to invade (see Figure 5.5). Each time the Saudi regime detected these tank movements, they called their "hot line to the Iraqi military command"[27] and received apologies and tank withdrawals. The third time, however, the Iraqis did not pick up the phone.

---

25. Statistics and discussion based on Williams 1999.

26. The account provided by Prince Bandar is drawn from his interview in Ottaway 1996.

27. Ottaway 1996, n.p.

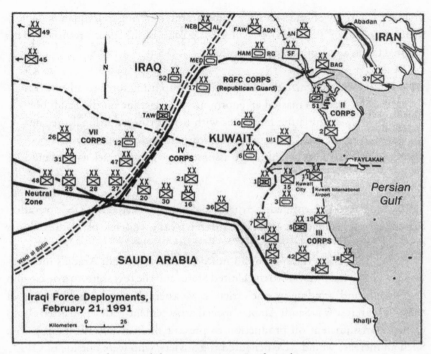

FIGURE 5.5. Iraqi forces and their proximity to Saudi Arabia.

Copyright: Dennis P. Mroczkowski, *U.S. Marines in the Persian Gulf, 1990–1991: With the 2nd Marine Division in Desert Shield and Desert Storm*, U.S. Marine Corps, History and Museums Division, Washington, DC, 1993.

Whether Bandar shared this information with President Bush during their private meeting is unclear, but Bandar recounts that Bush's commitment was clear, claiming that the president "extended his hand to him and said 'If you ask for help from the United States, we will go all the way with you.'"[28]

President Bush held to his promise, and on August 8 he addressed the country and the world through a televised speech to make his case for defending Saudi Arabia. In this speech, Bush immediately focused on the threat to Saudi territory before referencing Kuwait: "At my direction, elements of the 82d Airborne Division as well as key units of the United States Air Force are arriving today to take up defensive positions in Saudi Arabia. I took this action to assist the Saudi Arabian Government in the defense of its homeland."[29] He then presented a

---

28. Ibid.

29. Bush 1990, n.p.

dramatic summary of Iraq's invasion into Kuwait, referring to Iraq's tank deployment as operating "in blitzkrieg fashion." But soon Bush's speech returned to his original focus, a fear of a second invasion:

> But we must recognize that Iraq may not stop using force to advance its ambitions. Iraq has massed an enormous war machine on the Saudi border capable of initiating hostilities with little or no additional preparation. Given the Iraqi government's history of aggression against its own citizens as well as its neighbors, to assume Iraq will not attack again would be unwise and unrealistic.[30]

Lastly, Bush starkly identified the threat and clearly asserted the US motivation to help, stating, "Let me be clear: The sovereign independence of Saudi Arabia is of vital interest to the United States."[31]

While the rest of the speech did not delineate *why* Saudi Arabia's independence was of "vital interest" to the United States and the text shifts to a discussion of principle and standing by one's friends, we assert that a major dimension of that vital interest was Saudi Arabia's special arrangement with the United States to maintain sufficient oil production to prevent dramatically rising oil prices. Such an increase would have benefited the Soviets, who were outside of OPEC but nevertheless driven by their role as an exporter in the European hydrocarbon market.[32] At the same time, rising oil prices would hurt importing states such as the United States, and Bush likely feared that Saddam Hussein would not pursue global pricing policies that were as congruent with American economic interests as those of Kuwait.

President Bush's thoughts on the threat to Saudi Arabia were widely shared in Western media. National news in the United States was replete with stories with lead sentences such as "The Iraqi army moved into position for a possible attack on Saudi Arabia on Friday after its rapid conquest of Kuwait."[33] Worldwide condemnation swiftly followed Iraq's invasion into Kuwait, and as pressure mounted, Iraq declared that it would prepare to withdraw from its neighbor. But it had already put new leadership in place and declared that Kuwait would become a part of Iraq. As Iraqi forces prepared to withdraw,

---

30. Ibid.

31. Ibid.

32. See Chapter 6 for a rich discussion of Russia and the European hydrocarbon market.

33. Multiple Wire Reports 1990, A1.

there were concerns that they were merely repositioning for the next invasion into Saudi Arabia. Friedman writes, "With Iraq having overrun Kuwait, the crisis in the Persian Gulf appears to be turning into a struggle between the United States and Iraq for influence over Saudi Arabia and its vast oil reserves."[34]

One last issue focuses our attention on Iraq's ultimate market power strategy. Given Iraq's relatively poor access to the Persian Gulf (especially during its war with Iran), roughly 90 percent of Iraq's oil exports flowed through pipelines through Turkey and Saudi Arabia to the Red Sea. Invading Saudi Arabia would not only remove the threat of disrupted exports, it would ensure that Hussein controlled the flow of oil from oilfield to port. Together, these factors comprise a set of incentives that point to a motivation to invade Saudi Arabia. Although the world condemned the first foray into Kuwait, the military act was hugely successful. Hussein knew that Saudi forces were better equipped but still did not match the strength of his army.

Without help from an actor like the United States, Saudi Arabia stood a good chance of losing large amounts of territory that was long regarded as "one of the richest oil deposits in the World."[35] Thus, the United States was motivated to defend Saudi Arabia in order to prevent Iraq from seizing this market power opportunity (see Figure 5.6). Despite its concerns, the United States did not immediately turn to the use of military force against Iraq. As we have noted, Bush's speech on August 8 announced the deployment of troops to Saudi Arabia. This was followed by the positioning of additional US forces at Saudi military bases and the waters of the Persian Gulf. These forces were joined by troops from North Atlantic Treaty Organization (NATO) allies and Arab countries. Eventually over thirty countries joined this US-led Coalition. Their mission, code-named Operation Desert Shield, was initially defensive in nature. By building up its forces in the region, the United States and its Coalition allies aimed to deter Iraq from pursuing an invasion of Saudi Arabia.

Maintaining the status quo, however, would not be sufficient to address the threat from Iraq. As long as Iraqi forces remained in Kuwait, the possibility of Iraq taking future action against Saudi Arabia to pursue a market power opportunity remained. To prevent this, the United States and its allies needed to compel Iraq to leave Kuwait. Beginning immediately after the invasion, a series of twelve UN Security Council resolutions condemned Iraq's actions and

---

34. Friedman 1990, 1.

35. El Ghoneimy 1966, 697.

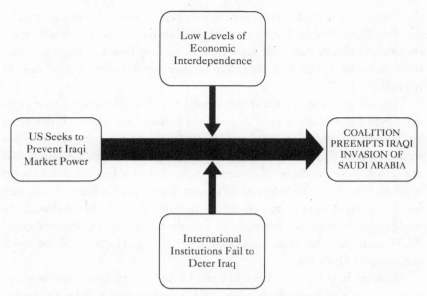

FIGURE 5.6. The Coalition's use of force to prevent Iraqi market power.

imposed economic sanctions on Iraq. The last of these—Resolution 678 passed on November 29—set a January 15 deadline for Iraq to comply, after which UN member states were authorized to "use all necessary means to uphold and implement" the previous resolutions and "to restore international peace and security in the area." However, none of these diplomatic activities persuaded Iraq to retreat from Kuwait.

When the January 15 deadline passed without any action by Iraq, the US-led Coalition eventually turned to the use of force to end the Iraqi occupation of Kuwait. Early in the morning of January 17, the Coalition initiated air strikes against Iraq. This strike commenced the transition from the largely defensive Operation Desert Shield to the offensive Operation Desert Storm. For over a month, the Coalition pursued an air campaign against Iraqi military targets in Kuwait and Iraq. Iraq responded with Scud missile attacks against Saudi Arabia and Israel and attempted an unsuccessful invasion of Saudi Arabia. However, the persistent Coalition air campaign severely damaged Iraq's military capability. On February 24, the Persian Gulf War entered a new phase as the Coalition launched a ground offensive into Kuwait. Within 100 hours, the Coalition forces had pushed Iraqi forces out of Kuwait and into southern Iraq. With the mission to "liberate" Kuwait completed, the Coalition ended its offensive, and a ceasefire was declared on February 28.

Market Power through Environmental Terrorism

Even as United States and Operation Desert Storm successfully compelled his forces back home, Saddam Hussein maintained his strategy of using force to revise Iraq's competitiveness in the oil market. Realizing that his attempt to increase his market power in the oil world had failed, Hussein employed a second strategy to improve his economic competitiveness by dismantling the Kuwaiti oil infrastructure. Iraq's withdrawal from Kuwait was so destructive that it caused scholars to coin the term "environmental terrorism."[36] The Iraqi military set fire to and damaged or destroyed over 750 oil installations in Kuwait, impairing almost all of its oil storage facilities, pipelines, and infrastructure, and crippled the Kuwaiti GDP to less than a third of its pre-war levels.[37] Oil was dumped into the Gulf and onto Kuwaiti soil, contaminating wells and desalinization facilities. Oil fires set by the Iraqis burned for nine months, and the subsequent health effects on the Kuwaiti population were devastating.[38]

This destruction was economic warfare by Iraq. While Hussein's first preference was to horizontally integrate this infrastructure and resources into the Iraqi economy, his second preference was to destroy it. Compared to leaving Kuwait intact, this middle option had the potential to improve Iraq's power in the oil market. The sanctions imposed by the international community in the wake of the war likely undid any of this advantage, but Hussein could not have predicted these costs accurately.

The destruction was terrible but also very effective in terms of achieving Iraq's aim of reducing global oil capacity. Hawley estimates that nearly 600 fires burned in Kuwait's oil wells, and the Kuwait Oil Company (KOC) declared 732 blown wells.[39] The KOC estimated in May 1991 that "the wells were burning 6 million barrels per day, 10 percent of the entire world's daily production."[40] Malice, no doubt, played a role in Hussein's decision to destroy Kuwait's oil wells in this fashion. Yet, his strategy was also consistent with one final effort to curtail Kuwait's flow of oil into the market. If Hussein had a chance at escaping his self-imposed situation intact and with a functioning state, he would have been able to count on no oil coming out of Kuwait.

---

36. Finnie 1992; Hawley 1992.

37. Casey 2007, 107.

38. Ibid., 108–109.

39. Hawley 1992, 14.

40. Ibid.

In sum, Saddam Hussein was motivated to increase Iraq's market power and gain the ability to control the global output and prices of oil. This provided an incentive for Iraq to seize control of oil resources in Kuwait and Saudi Arabia to increase its share of the market. However, according to our model of market power politics, such a motive is not sufficient to lead to a strategy of violence. To understand why Iraq opted to use force to pursue this market power opportunity, we also need to consider the potential constraints provided by economic interdependence and international institutions. As we will show in following sections, these constraints were fairly weak and provided little to deter Iraq from turning to military action.

Economic Interdependence: No Leverage against Iraq

Iraq faced few economic constraints on using military force to pursue its market power objectives. In 1990, there was a low level of economic interdependence between Iraq and Kuwait. Despite being neighbors, Iraq and Kuwait were not active trading partners. Their most intense economic ties came in the form of loans from Kuwait to Iraq during Iraq's war with Iran. In terms of goods and services, however, the two states had little to offer one another. Examining the bilateral exchange of imports and exports for Iraq during 1985–1989, for example, shows that Kuwait was not a top importer from or exporter to Iraq. Iraq's top sources for imports in 1989 were the United States, Germany, Great Britain, Japan, and France. Its top five export targets were the United States, Brazil, Turkey, Japan, and France.[41]

Perhaps more importantly, the two states were importing and exporting the same goods with other states. Both states produced oil, of course, and used the revenue from oil sales to purchase the goods and services they needed from other states. As such, they were economic competitors, even though they shared membership in the OPEC cartel. As we will explain in the next section, the OPEC regime did not have the structure necessary to overcome the competitive quality of their economies.

As such, Kuwait did not have the ability to leverage its economic ties with Iraq to deter Saddam Hussein from action. Kuwait could not effectively sanction Iraq, nor could it benefit from the threat to cease buying or selling goods and services. Kuwait's only leverage lay in its oil production decisions, but it would have been nearly impossible to credibly commit to ceasing oil production in order to

---

41. Gleditsch 2002.

allow Iraq to benefit from increased prices. Even if Kuwait seriously considered such a strategy, other OPEC and extra-OPEC producers could step up production in an attempt to enjoy the increase in prices that would result from a drop in Kuwaiti production, thus negating its effects. Moreover, once Iraq had removed the threat of force, there would be little pressure remaining on Kuwait to maintain an artificially low output. This time-inconsistency problem made it impossible for Kuwait to commit to a solution that would satisfy Iraq.

Turning to Saudi Arabia, the leverage of economic interdependence is slightly more complicated. The two states did not trade much with one another and were similarly aligned as economic competitors, much like the Iraq-Kuwait dyad. Saudi Arabia, however, had two mechanisms by which it could put economic pressure on Iraq. First, it could increase its own oil exports as OPEC's swing producer to negate any price increase resulting from the cessation of Kuwait's production.[42] Second, as we mentioned earlier, Iraq funneled much of its oil exports via a pipeline running through Saudi Arabia on the way to the Red Sea. Saudi Arabia could have shut down the pipeline, forcing it to export its oil through pipelines to Turkey or via Gulf ports.

We believe this was a less effective threat than the Saudis hoped for two reasons. First, with the absorption of Kuwait, including the Bubiyan and Warbah islands, Iraq had improved access to exporting oil through the Gulf. Second, and perhaps more ominously, an Iraqi invasion into Saudi Arabia would remove this leverage permanently.[43] Thus, while the pipeline through Saudi territory would constitute a degree of Iraqi dependence in a normal economic environment, the invasion into Kuwait (and potential invasion into Saudi Arabia) negated that dependence completely.

Finally, not even the major Western powers such as the United States, Germany, or the United Kingdom could leverage their economic ties to Iraq to dissuade it from invading Kuwait. Given the large oil reserves in Iraq and Western dependence on oil, the asymmetry in the economic interdependence between Iraq and the West favored Iraq. Indeed, if Saudi Arabia had not pledged to increase oil production to offset the drop in Iraqi and Kuwaiti exports, the United States may have had a harder time mobilizing support for a counter-invasion. Even military arms sales from the United States, which were significant,

---

42. This is indeed what Saudi Arabia did once the United States and the Coalition launched Operation Desert Storm. Saudi Arabia increased oil production to prevent a shortage of oil in the market.

43. Which it did, in a way. The Iraq-Saudi Arabia oil pipeline ceased to be in use after Hussein's invasion into Kuwait in 1990. Subsequently, Iraq has primarily exported its oil through pipelines into Turkey and Syria, as well as through the Gulf.

represented a symmetrical linkage that the Americans would struggle to leverage. The United States sought to use its ties to the Iraqi military and its history of weapons sales to Iraq, but with the end of Iraq's war with Iran, Saddam Hussein's need for additional equipment was diminished.

Saudi Arabia's pledge was no doubt contingent on the willingness of the United States to offer protection and to expel Hussein's army from Kuwaiti soil. Once the Saudis made good on their promise, Iraq was similarly stymied in its hopes of leveraging its oil power to stop the Coalition from a counter-invasion. In essence, the Saudis eliminated any economic advantage Iraq may have had over the West by using its swing-producer capability to prevent a shortage in the oil market. Their increased production made it impossible for Saddam Hussein to constrict the Middle Eastern oil market in a bid to thwart the counter-invasion.

International Institutions: Insufficient Solutions Unable to Prevent War

Given the low level of economic interdependence, international institutions faced a greater hurdle to constrain Iraq's motivations to use force to acquire greater market power. Here we consider three sets of potential institutional constraints and find that in each case, institutions had little ability rein in Iraq. First, OPEC did not provide a viable avenue for Iraq to gain increased revenue through higher oil prices. Second, Iraq had a weak legal claim to Kuwaiti (or Saudi) territory under international law. Finally, organizations like the Arab League and the United Nations did not provide procedures that could effectively prevent war in this case.

OPEC

The cartel-style regime of OPEC failed to provide an alternative nonviolent institutional mechanism by which Iraq could achieve its revenue goals. This was partly due to the mismatch between the incentives for Iraq and the broader goals of OPEC. The essential challenge for OPEC management was the classic problem of individual short-term gains at odds with the market focus of the organization. While the cartel was concerned with the price of oil, individual OPEC members wanted to maximize their own revenues. These two dual goals were hard to recognize. As Mabro explains, "The price that maximizes the revenues of a small producer does not always suit the revenue objective of the price makers in the short/medium run, and rarely, if ever, in the longer term."[44] The structure of the

---

44. Mabro 1992, 3.

cartel made OPEC predisposed to defections by members who sought to assuage individual needs at the expense of the broader goal.

The tension between individual member needs and OPEC goals becomes particularly salient when exogenous shocks exacerbate the revenue needs for governments. The 1970s brought an increase in power to the bloc of oil states as a result of strong member coordination to mitigate competition from states outside of the cartel. But the market adapted rapidly, expanding extra-OPEC production to counter OPEC's control or just to take advantage of high oil prices. Mabro demonstrates that during the years before Iraq's invasion, OPEC was unable to exercise much control over oil prices despite numerous attempts:

> In certain circumstances the core producers may find themselves con-
> fronted with a big increase in supplies from the fringe largely caused by
> exogenous factors. And if this happens to coincide with a downward shift
> in the demand curve due to a severe recession, to significant conservation
> induced by policy, technical progress or other phenomena, the core pro-
> ducers may be left with a very small volume of residual demand. This is
> precisely what happened in 1981–87.[45]

In the aggregate, these increases in supply from small states add up, preventing core producers from leveraging demand to increase prices.

Thus, by the mid-1980s, Saudi Arabia recognized that extra-OPEC competi-
tion had diminished OPEC's overall market share, and as a result, OPEC's ability to achieve its market goals. To counter these competitors and improve OPEC's market power, Saudi Arabia increased production in an effort to regain OPEC's overall market share of oil exports. The effort was successful, but the result was a nearly 55 percent drop in prices between 1985 and 1986.[46] Saudi Arabia's role as the swing producer within the cartel was solidified, but states like Iran and Iraq became desperate for changes that would improve their economy and ability to pay for their war with one another.[47]

By the time Saddam Hussein exited his war with Iran, OPEC was unable to provide him with the opportunity to spike oil prices even if he could have

---

45. Ibid., 5.

46. Almoguera, Douglas, and Herrera 2011, 163.

47. Saudi Arabia began functioning as a swing producer after the March 1983 OPEC meeting, whereby it increased or decreased production to enable other member states to maintain their full quotas. See ibid., 162–163.

convinced key players to accommodate him. The dual role the Saudis played—swing producer and overall market manipulator—prevented Iraq from raising oil revenues quickly through increased production. Without the ability to constrain the price of oil or the output of other major suppliers, Iraq's own increases in production would only add to an already flooded market and decrease oil prices. The resulting price drop would undermine Iraq's primary goal of increased revenue.

Iraq tried to lean on other states to create room for increased production without decreased prices, but the OPEC institution was not well suited for inter-party enforcement. OPEC was primarily designed to promote collusion among member states but has always faced the challenge of enforcement when individual states exceed their allotted quotas of oil production. OPEC does not contain enforcement mechanisms for its policy decisions. As a cartel, OPEC's efficacy for its members emerges from its ability to provide them with information such that cooperative joint policies regarding production and price targets can be obtained. But as individual members diverge from this collective policy, the other states have little or no ability to coerce compliance. If OPEC had contained language that would have enabled meaningful punishments for states that violated quota agreements, Iraq might have been able to leverage the regime to rein in Kuwaiti and UAE oil production without moving troops. However, without such enforcement capability, OPEC was not able to provide Iraq with a non-violent institutional settlement compatible with its market power goals.

International Law

Second, Iraq had little ability to achieve its territorial expansion goals under international law. Iraq had maintained a long-standing territorial claim that Kuwait should be part of the Iraqi province of Basra. However, this legal claim was rather weak. The claim was tied to pre–World War I conditions, when both Iraq and Kuwait were within the Ottoman Empire. Subsequent to the war, both nations came under British control. Iraq was granted statehood by the League of Nations in 1932, but Kuwait did not seek independence from the British until 1961. Throughout the pre–World War I period, Kuwait's ties to Basra were simply tied to the fact that Basra housed the regional Ottoman authorities. Iraq may have never agreed to a formal demarcation of the border, but its protest of the legalities of the state of Kuwait has been summarily rejected.[48]

During the negotiations at Uqair in 1922, negotiations favored Iraq in part because they were being facilitated with a heavy hand by Sir Percy Cox. Cox was

---

48. Pillai and Kumar 1962; Greenwood 1991.

the British High Commissioner to Iraq and outranked the corresponding British representative supporting Kuwait. In our investigations of the Uqair Protocol, we find that it was Kuwait who was most upset about the boundaries being drawn by Cox. The House of Saud, which ruled the Najd (the future Saudi Arabia), protested the establishment of formal borders at all, given the needs and traditions of nomadic tribal movements and the inability to observe any sort of natural boundary between Iraq and the Najd. Kuwait, for its part, was upset about the large "neutral zone" established between Kuwait and what would become Saudi Arabia. That dispute concerned the southern Kuwaiti border, however, and was independent from its border with Iraq. Lastly, upon achieving independence from the British, Kuwait was accepted into the United Nations and the Arab League and has since been a voting member of the UN General Assembly. Regardless of Iraq's assertions, the global community considered the legality of Kuwait to be settled law. Thus, Iraq did not have the possibility to pursue international legal procedures to expand its territorial reach over Kuwait.

The Arab League and the United Nations

Finally, the design of regional and global organizations failed to provide procedures to prevent the use of force. Consider the case of the Arab League. Despite an explicit prohibition of violence against other member states in the Charter and a declaration that decisions by the Council of the Arab League (representatives from each member state) are binding, the Charter requires unanimity within the Council to make such decisions. Pinfari contends that this structural explanation is exaggerated and that the Arab League indeed made a good faith effort to mediate the dispute to avoid war. Instead, his analysis reveals that the Arab League was more successful in supporting member states in conflicts against non-member states, particularly Israel.[49] The Arab League did convene a summit on August 10, roughly one week *after* the invasion. The resolution passed at the summit called for the immediate withdrawal of Iraqi forces from Kuwait. It is telling, perhaps, that the previous summit, held in May earlier the same year, focused almost entirely on immigration from the Soviet Union into Israel.[50]

Like the Arab League, the United Nations was designed to respond to but not prevent the attack itself. The UN Security Council passed twelve resolutions concerning Iraq and Kuwait in 1990.[51] The first resolution, however, comes on

49. Pinfari 2009.

50. Gunay 2000.

51. United Nations Security Council 2020.

August 2, *after* Iraq invaded Kuwait. Once again, the problem is one of structure, in that the Security Council is institutionally designed to respond to crises rather than to prevent them. Moreover, once it did take action, the United Nations was unsuccessful in its attempts to use nonviolent methods to compel Iraq to withdraw from Kuwait. Instead, when Iraq ignored the UN's January 15 deadline to comply with the Security Council resolutions, the US-led Coalition had to turn to military force to end the occupation.

Putting these economic and institutional components together, neither of the two sources of friction were sufficient to slow down or deter Saddam Hussein from his attempt to absorb Kuwait as the nineteenth district of Iraq. Nor, do we expect, would they have been able to deter Saddam Hussein from a rapid strike into Saudi Arabia following Iraq's initial success in Kuwait. Given the imbalance between motive and constraint, Iraq turned to violence to pursue its market power goals, while the US-led Coalition responded with military action to prevent a significant shift in market power. Overall, then, it seems plausible to conclude that it was the market power opportunity to control the global output and price of oil that drove Saddam Hussein and the state of Iraq to war with Kuwait. In the next section, we turn to a series of alternative explanations to gauge the extent to which our market power explanation holds up to other accounts of the causes of the invasion of Kuwait and the Persian Gulf War.

Alternative Explanations

There are three alternative explanations to consider when assessing the motivations of the Iraqi regime and its decision to invade Kuwait: Iraq's territorial claim to Kuwait; that Iraq simply wanted more oil, not control over the price; and that Iraq desired Kuwait for expanded access to the Persian Gulf. We attend to each possible alternative here and provide evidence that the first two alternatives do not hold up to empirical scrutiny. The third alternative, that Iraq desired better access to the Persian Gulf, was more likely to have contributed to Hussein's calculations when launching his invasion, but was also ultimately a function of the underlying desire to control Kuwait's oil production and export capacity.

Iraq's Territorial Claim on Kuwait

The first alternative rationale for the invasion is Iraq's long-standing territorial claim on Kuwait. Iraq has maintained since Kuwait's entrance into the international system in 1961 that Kuwait is rightfully a southern part of the Basra province of Iraq. Moreover, the 1990 invasion was not Iraq's first attempt to use force to challenge Kuwait's statehood. Beyond its initial protest in 1961, Iraq initiated several border incidents. The most intense incident involved the massing of troops at the border in 1973. The Iran-Iraq war altered Iraq's focus, however, and Kuwait allied with Iraq during this war.[52]

There are three reasons in particular to doubt the validity of Iraq's territorial claim and its impetus for Iraq's decision to invade in 1991. First, as we discussed earlier, the legal claim itself is rather weak.[53] Second, Iraq's territorial claims over Kuwait did not begin until *after* oil was discovered in Kuwait. Apparently, the desire to make Kuwait a part of Iraq was less appealing when Kuwait's main industry was pearl harvesting. The newfound wealth fed the appetite for horizontal integration. Third, Iraq backed off of its territorial claim during the Iran-Iraq war and enjoyed Kuwaiti assistance throughout the war. Only in the wake of the exhaustive war did Iraq return its focus to the border dispute. Iraq's tacit agreement on its borders with Kuwait during the Iran-Iraq war stood in stark contrast to Hussein's demands in July 1990, and this shift in stance is plausibly explained by Iraq's need for revenue following the war.

Iraqi Oil Grab

One of the most pervasive alternative explanations to our interpretation of Iraq's invasion as a quest for market power is the notion that it was simply after more oil and better access to the Persian Gulf. It is important to note, however, that Iraq had as much oil as it could produce (and more). Limits to Iraqi oil production were self-induced and in coordination with the OPEC cartel. Perhaps most importantly, petro-experts at the time read the situation in the same way that we do now. Consider the following interview of Pierre Terzian, who in 1990 was the head of the Petrostrategies consulting firm:[54]

Editors: At the time of the Iraqi invasion, many people said that Iraq wanted to control all the oil reserves in the Gulf. How do you react to that hypothesis?

Terzian: I don't believe it holds up. Iraq has no need of other oil reserves; its own reserves are very large and its potential production capacity is far from being exhausted. And even though it has a relatively long oil history,

---

52. Casey 2007; Finnie 1992.

53. Greenwood 1991; Pillai and Kumar 1962.

54. Terzian 1991, 102–103.

Iraq is still for the most part practically unexplored for oil—only a very small part of Iraqi territory has been explored relative to its potential. The Iraqis really don't need the oil of neighboring countries. If they had, or have, such ambitions, in my opinion they would be absolutely senseless.

Editors: What about in terms of controlling the market?

Terzian: That's another issue. What Iraq needs is not more reserves, but oil prices that would allow it to meet its financial needs—to import goods, to pay off the enormous debt contracted during the Iran-Iraq War, and finally to boost an economy that suffered greatly during the war.

This discussion clearly separates the motivation for increased supply from the motivation to control the market. From Terzian's description, Iraq's problem was not that it lacked oil; it had as much as it wanted and then some.

Indeed, Iraq was the OPEC whip in terms of pushing states to decrease output in 1989 and early 1990. As major players like Saudi Arabia agreed to begin reining in production since OPEC's market share had increased, Iraq grew increasingly frustrated with overproduction by Kuwait. As Terzian states, "The tension reached a peak last spring [1990] when, because of this overproduction by Kuwait and the Emirates, the prices fell by almost 50 percent within a few weeks. Prices had been $20, $21, even $22 a barrel in January, and dropped to $13 or $14 in the spring."[55] Just pushing out more oil would not have solved Iraq's cash problem, even if prices were stable. Indeed, it would have made things worse by causing a further decline in prices. It is important to note that without price-setting capability, an increase in overall oil production spurred by Iraq would have decreased the world price. Such a strategy would have undermined the primary goal of the invasion, which was to refill the government coffers that were so depleted by the previous war. Hussein's only option to solve his economic woes from the Iraq-Iran War was to set the price, which meant holding his production steady while decreasing everyone else's.

The same can be said about the US response. If the White House could have been assured that Saddam Hussein would stop with Kuwait and the world was convinced that Saudi Arabia was safe from Iraqi forces, the West may never have resorted to force to remove Iraq from Kuwait. Thus, to the frequent refrain that the two modern wars between the United States and Iraq (1990 and 2003) were about oil, we partially disagree. With respect to the first event, the Persian Gulf War, the conflict was about the *price* of oil and the ability to manipulate that price.

---

55. Ibid., 101.

Improved Access to the Gulf

A more challenging counter-explanation is that the Iraqi regime sought improved access to the Persian Gulf for exporting its own oil. This challenge to our interpretation of the 1990 invasion is that Iraq simply changed its strategy from seeking broader control of the Shatt al-Arab to a new strategy of using Kuwait's Bubiyan and Warbah islands and the Port of Shuwaikh to improve its access to the Persian Gulf.[56] The dispute between Iran and Iraq over the Shatt-al-Arab was one of the primary motivations for that war, and the resolution of the war left Iraq with no improvement over its control of the river and its oil routes. Without controlling Kuwait, the legal interpretations of access to the Gulf largely left Iraq with no options other than the Shatt al-Arab. During the Iran-Iraq war, Iraq's oil exports were almost completely transported by pipelines through Saudi Arabia and Turkey. Controlling the Bubiyan and Warbah islands would radically reshape Iraq's legal access to the Gulf according to the United Nations Convention on the Law of the Sea.

These goals may have been part of the Iraqi motivation, but they are in essence identical to the motivation to integrate Kuwaiti oil. Even if they were a priori goals, however, they did not appear as major issues in the pre-war negotiations between Iraq and Kuwait that took place in Saudi Arabia and Baghdad in 1990.[57] Moreover, the desire for better access to the Gulf coincides with the price-control goal established earlier. Improved Gulf access would have enhanced Iraq's ability to use the Gulf for oil exports and reduce its reliance on the pipeline through Saudi Arabia to export oil from its southern wells.

Overall, perhaps the best interpretation of this alternative explanation is that it is not incompatible with our own. Both motivations seem to have been in play for Iraq in the summer of 1990, and the market power explanation does not fully subsume the pursuit of better port access to the Persian Gulf. This preference for improved access, however, is not historically referenced as the key demand made by Iraq. With respect to this last alternative explanation, we believe it coexists nicely with the market power explanation we set forth herein.

Conclusion

The Iraqi invasion into Kuwait surprised much of the world, and the Coalition's military response to defend Saudi Arabia and push Iraq out of Kuwait must

---

56. Anscombe 1997; Crystal 1995; Casey 2007.

57. Casey 2007.

have surprised Saddam Hussein. From our perspective, this case illustrates how unchecked motivation, driven by potential market power opportunities, can heighten the probability of war. Hussein's desire to gain the ability to affect oil prices to increase revenue provided a motivation for his decision to invade Kuwait. Relatively unconstrained by low levels of economic dependence and a lack of acceptable institutional solutions, he was willing to turn to violence to pursue these market power goals. Following the invasion of Kuwait, many feared that a subsequent invasion of Saudi Arabia, the world's largest oil producer, would soon follow. To prevent such a shift in market power, the United States sent troops to Saudi Arabia and formed an international coalition that eventually attacked Iraqi forces in Kuwait.

Ultimately, the effort led by the United States to push Iraq out of Kuwait and to prevent a subsequent invasion of Saudi Arabia represents a major turning point in world history. While the initial mission to push Iraq out of Kuwait was successful, the world continues to grapple with the unintended effects of the Persian Gulf War. For the United States, the initial military victory over Iraqi forces was dramatic and extraordinary. American civil society celebrated the decisive win, seemingly exorcising the ghosts lingering from the Vietnam War. Additionally, the success of the UN-backed Coalition signaled the possibility that, in the immediate aftermath of the end of the Cold War, the UN Security Council could play a more active role in punishing rogue behavior in the international community.

Long-term consequences, however, turned out to be much more troubling. The decision to leave Saddam Hussein in power and avoid pushing Coalition forces all the way into Baghdad seemed generous and wise at the time, and it was perhaps vindicated years later by the counterfactual that was enacted in the 2003 Iraq War. At the same time, however, Hussein's survival (both literal and political) enabled him to engage in terrible reprisals against Shi'a Muslim and Kurdish communities in Iraq as a tactic to hold on to power. The resulting No-Fly Zones imposed by the United States and its allies following Iraq's defeat would become a semi-permanent but frustratingly limited solution to the problem of protecting these communities and preventing Hussein from moving against his foes. The result was a decade of stalemate that weakened Hussein's grasp on power, but at great cost to civilians and infrastructure.

Other unexpected consequences continue to impact world politics today. The presence of American troops on Saudi soil served as motivation for radical Islamic groups such as Al Qaeda, fed by the resources of Osama bin Laden, who was adamantly opposed to foreign troops in Saudi Arabia. The presence of "infidels" on holy ground became a rallying cry, fueling a new generation of violence against governments and civilians all over the world. Al Qaeda's assault on the World Trade Center in New York and on the Pentagon, near Washington, DC, on September 11, 2001, is the most infamous of these terrorist attacks, killing thousands of American citizens and triggering a war on terror that continues today.

Beyond the military response that has characterized American foreign policy for the last two decades, the effects of the Persian Gulf war on oil politics have also been powerful and enduring. Saudi Arabia and the United States continue to maintain a strategic alliance, fed by the joint need to manage the global oil market. American investment in domestic oil and gas production ramped up, in part based on the motivation to be more self-sufficient as a national security priority, resulting in more than double the production of petroleum products and a 600 percent increase in exporting hydrocarbons.[58] While this level of production is unlikely to last long, it has generated lasting effects for the global market and slowed the transition to green energy alternatives.

Stepping back to the analytical lens of the book, this chapter highlights the path to war that can result from market power politics. For Iraq and Saddam Hussein, the potential benefits from market power were too tempting. For the United States and Western Europe, the risk of Iraqi price-setting power was too great, and the resulting war between the Coalition and Iraq is a sobering example of how even complex, advanced economies such as those in the Organisation for Economic Co-operation and Development (OECD) can be vulnerable to the consequences of market power politics.

Although the events discussed in this chapter transpired three decades ago, the global oil and natural gas markets remain a constant source of concern for the world economy. In the spring of 2020, Saudi Arabia and Russia entered into an oil price war after Russia opted not to adhere to new OPEC resolutions to reign in oil production. While not a formal member of OPEC, Russia is among the oil-exporting states that along with the formal members represent OPEC Plus. As the United States ramped up production, OPEC Plus states had often worked together to manage their own production with the goal of stable prices. In early March 2020, OPEC states agreed to additional cuts in production, but Russia quickly declined to participate in the cuts.

Why was Russia so reluctant to cut its production in coordination with OPEC? Perhaps Russia saw an opportunity to increase sales and market share as the core OPEC states reigned in production. The resulting price war led to a precipitous drop in oil and natural gas prices, triggering some of the lowest

---

58. US Energy Information Association 2019.

prices for these commodities in decades. In the next chapter, we examine the importance of hydrocarbon exports for Russia. We will attempt to demonstrate that these exports are not only an essential dimension of the Russian economy, they also represent a market power opportunity that Russia is willing to fight over.

6

Russia

CORNERING THE GAS MARKET

IN CONTRAST TO Iraq's violent land grab of Kuwait in the summer of 1990, Russia has used less overt tactics to expand its control into neighboring territories in Georgia and Ukraine in recent years. In Georgia, the perpetual "frozen conflicts" in South Ossetia and Abkhazia have allowed Russia to slowly consolidate its power in these breakaway regions through salami tactics. While the official status of these regions remains disputed, Russia has moved step by step to integrate South Ossetia and Abkhazia militarily and economically without formally annexing them. In recent years, Russia has even pursued a "borderization" strategy in which it has physically moved boundary posts in order to expand South Ossetia's territory. With this "creeping occupation," Russia aims to expand its de facto control further into Georgia.

In Ukraine, Russia took a more extreme tack when it moved to formally annex Crimea in 2014. However, unlike Iraq's armed invasion of Kuwait, Russia mainly relied upon gray zone tactics to absorb Crimea into the Russian Federation. Taking advantage of political uncertainty in the midst of the Euromaidan protests, non-uniformed Russian soldiers laid the groundwork for Russia's eventual takeover of Crimea. With their help, Russia moved arms and military equipment onto the peninsula and pressed Crimean officials to hold a quick referendum that resulted in Russian annexation of Crimea. By using these "little green men," Russia was able to pursue its territorial ambitions without resorting to open warfare. Along with its activities in Crimea, Russia has also provided support to separatist forces in the eastern Ukrainian regions of Donetsk and Luhansk. The emergence of potential "frozen conflicts" in these regions provides Russia with an opportunity to follow a similar pattern of control to the one it has pursued in South Ossetia and Abkhazia.

In this chapter, we examine how Russia's pursuit of these gray zone tactics in Georgia and Ukraine can be seen as part of its overall strategy to preserve and expand its market power in the natural gas market. Unlike Iraq, which turned to the use of force in an attempt to become the dominant player in a hydrocarbon market, Russia has already achieved a privileged position in the European gas market. Given its significant reserves and existing pipeline infrastructure, Russia is the predominant gas supplier to many European countries. This gives Russia and its state-owned gas company, Gazprom, potential price-setting capabilities, which can be used to extract rents or exert political leverage.

Several ongoing challenges, however, threaten Russia's market power position. For example, transit states have the potential to disrupt the flow of gas from Russia to customers, reducing Russia's ability to leverage its market power. At the same time, importing countries have incentives to seek potential alternative suppliers of natural gas. In recent years, Russia has faced increased competition in Europe from other pipeline suppliers and producers of liquified natural gas (LNG). Additionally, changes in the structure of the gas market—including an increased use of hub pricing and liberalization efforts by the European Union— potentially threaten Russia's ability to set prices or use the provision of gas as a coercive tool.

Given these challenges, Russia must continuously pursue a set of strategies to preserve and potentially enhance its market power status. This strategic portfolio aims to control as much of the supply chain as possible, hinder rival gas suppliers, and expand the size of Russia's gas reserves. Some of these strategies are non-territorial in nature, such as vertically integrating downstream gas pipelines into Gazprom and building new pipelines that bypass transit countries. Gazprom's attempts at transnational vertical integration have had mixed success. The Russian firm has gained increased control of gas companies and pipelines in Moldova, Belarus, and Armenia. However, its efforts in Georgia and Ukraine have failed. Similarly, Gazprom has had some success in building bypass pipelines such as the Nord Stream pipeline, but it has often run into bureaucratic hurdles that have required it to delay or abandon pipeline strategies.

Other market power strategies directly revolve around disputes over territory. Such strategies become particularly useful when non-territorial strategies prove to be unsuccessful. Here we focus on the market power motivations behind Russia's territorial aggression against two of its neighbors, Georgia and Ukraine. Georgia is a key transit corridor for hydrocarbons from Azerbaijan to the west. By destabilizing Georgia, Russia could disrupt the supply of gas from a rival producer to Europe. On the other hand, Ukraine has been a key transit state for Russian gas. Annexing Crimea and supporting secession movements in eastern Ukraine provide Russia with an opportunity to gain access to additional gas reserves and increase its bargaining leverage over a transit state.

In these cases, Russia has largely relied upon strategic delay and gray zone tactics to pursue its territorial ambitions in Georgia and Ukraine. The high level of interdependence between Russia and EU countries constrains Russia from pursuing significant levels of military force. At the same time, international law does not provide Russia with many viable options for satisfactory institutional solutions that would be in line with its market power goals in Georgia and Ukraine. Given this configuration of economic and institutional constraints, Russia has eschewed both military escalation and institutional settlement. Instead, in line with the expectations of our theory of market power politics, Russia has turned to a strategy of delay, which allows it to achieve its goals over time through gray zone tactics.

The chapter proceeds as follows. First, to orient readers unfamiliar with the political economy of natural gas markets, we provide a snapshot of Russia's gas trade with Europe and describe the factors that have historically led to Russia's significant market power in the regional gas market. Next we outline a set of ongoing challenges to Russia's market power and discuss the portfolio of strategies that Russia uses to preserve and enhance its price-setting capabilities. We then discuss how our model of market power politics can help explain Russia's use of strategic delay and gray zone tactics in Georgia and Ukraine. Lastly, we consider some competing explanations for Russia's aggressive tactics toward its neighbors.

Russia and the European Gas Market

Russia is a significant player in transnational hydrocarbon markets. It accounts for about 18 percent of global gas production and 12 percent of global oil production.[1] Additionally, Russia is the largest exporter of natural gas and the second-largest exporter of crude oil in the world.[2] The Russian economy largely depends upon the production of these hydrocarbons, as oil and gas account for the majority of Russia's exports. Critically for the Russian government, over 40 percent of its federal government revenues came from oil and gas.[3] This reliance on

---

1. World Bank 2019.

2. BP 2019.

3. World Bank 2019.

hydrocarbons is likely to continue in the foreseeable future, as Russia holds the largest proven gas reserves in the world and the sixth largest proven oil reserves.[4]

While Russia is a significant producer of both oil and gas, this chapter will primarily focus on the politics of Russian activities in the natural gas market. Even though oil exports provide a greater source of income for the Russian government than gas imports, Russia enjoys a more privileged position in the gas market. Russia is a significant supplier of natural gas to many European countries, and in several cases, Russia's state-owned gas company, Gazprom, has or had near-monopoly status. Additionally, natural gas markets have historically been primarily regional, in contrast to more global oil markets. Given this, Russia has faced fewer potential competitors to its gas exports than to its oil exports. However, emerging shifts in the structure of gas markets potentially threaten Russia's ability to maintain its power in the European gas market. Thus, the gas market provides a good avenue to explore how market power opportunities shape the foreign policy strategies pursued by Russia.

To provide a picture of Russia's position in the gas market, let us take a look at international gas flows in 2017.[5] In that year, over 90 percent of Russian gas exports were transported through pipelines to European countries (including Turkey) and former Soviet republics in the Caucasus and Central Asia. Table 6.1 lists the top ten destinations for Russian pipeline gas in 2017. Over one-fifth of Russian pipeline gas exports went to Germany. Other large markets in the European Union for Russian gas are Italy, Slovakia, France, and Poland. Outside of the European Union, the largest importers of Russian gas are Turkey and Belarus. Overall, almost three-quarters of Russia's pipeline gas exports went to EU countries in 2017. In addition to pipeline gas, Russia exported a relatively small amount of LNG, primarily to East Asia. However, LNG constituted less than 7 percent of Russian gas exports in 2017.

Just as Europe is by far Russia's largest export market for gas, Russia is the largest external supplier of gas to Europe. In 2017, Russia supplied about 43 percent of pipeline gas imports into EU countries and 37 percent of their total gas imports (both pipeline and LNG).[6] However, the level of dependence on Russia varies among EU states. Table 6.2 shows the percentage of pipeline gas imports that came from Russia for several EU countries. Many central and eastern European states, including Austria, Hungary, Slovakia, and Finland, only imported pipeline gas from Russia. For other central and eastern European

---

4. BP 2019.

5. Gas trade data are taken from BP 2018.

6. BP 2018.

Table 6.1. Top Ten Importers of Russian Pipeline Gas, 2017

Country	Pipeline Gas Imports from Russia (billion cubic meters)	% of Russian Exports
Germany	48.5	22.5
Turkey	27.6	12.8
Italy	22.3	10.4
Belarus	17.8	8.3
Slovakia	13.7	6.4
France	11.5	5.3
Poland	11.3	5.2
Austria	8.6	4.0
Netherlands	8.6	4.0
Hungary	8.2	3.8

Source: BP 2018.

Table 6.2. Percentage of Pipeline Gas Imports Coming from Russia, 2017

Country	% of Imports
Austria	100.0
Finland	100.0
Hungary	100.0
Slovakia	100.0
Greece	81.9
Poland	76.0
Czech Republic	64.1
Germany	51.2
Italy	41.5
France	34.3
Netherlands	20.9
United Kingdom	10.2
Belgium	0.0
Ireland	0.0
Spain	0.0

Source: BP 2018.

states, including Germany and Poland, more than half of their gas imports came from Russia. Western EU countries, such as France, the Netherlands, and the United Kingdom, imported a smaller proportion of gas from Russia, and several western European states—including Spain, Belgium, and Ireland—did not import any Russian pipeline gas.

This variation in dependence upon Russian gas imports is largely a function of geography and technology. Some European countries, such as Norway, the United Kingdom, the Netherlands, and Romania, have domestic resources to meet a portion of their gas needs.[7] However, the vast majority of countries cannot rely upon domestic production and instead must import some or all of the gas they consume. Most of this imported gas is transported over long distances via pipelines. Construction of these pipelines requires significant time and capital, so in the short to medium term, gas consumers are largely dependent upon the existing pipeline network. For many countries in central and eastern Europe, most of these pipelines originate in Russia. Thus, they generally have to rely upon Russia to supply their gas imports. On the other hand, many countries in northern, western, and southern Europe have access to pipelines from alternative suppliers, including Norway, the United Kingdom, the Netherlands, and Algeria. For this reason, they tend to be less dependent upon Russian gas.

Thus, when we consider the European gas trade from the perspectives of both the exporter and the importer, we see much variation in bilateral relationships between Russia and European countries. Russian relations with many countries in eastern and central Europe are largely characterized by asymmetric interdependence. For example, the Baltic states, Hungary, and Bulgaria largely rely upon Russian gas and do not have many alternative foreign or domestic sources of natural gas. On the other hand, each of these countries receives a small fraction of the gas exported by Russia. In contrast, Russia's relationships with western European countries like Germany, Italy, and France are more symmetrically interdependent. These countries are among the largest importers of Russian gas and provide a significant fraction of Gazprom's revenue from exports. Since it would be difficult for Russia to make up these revenues elsewhere, Russia is more dependent upon exports to these countries. Additionally, compared to many of their neighbors to the east, western European countries have more balanced portfolios of gas imports and are thus less reliant on Russian gas. As we will see, this diversity in dependency relationships has shaped Russian foreign policy toward Europe and has complicated the European Union's response to Russia's attempts to maximize its power in the European gas market.

---

7. Grigas 2017, 141.

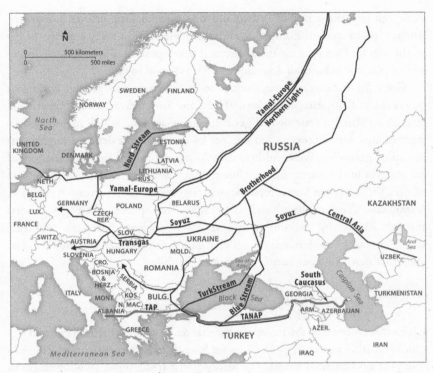

FIGURE 6.1. Russian gas pipelines to Europe.

The Emergence of Russian Market Power

The gas pipeline network connecting Russia to European consumers dates to the Soviet era (see Figure 6.1).[8] Initially, the Soviet Union began constructing pipelines to transport natural gas from fields in the Ukraine to satellite states in eastern Europe. Then, beginning in the 1960s and 1970s, the Soviets began making agreements with other European countries, including Austria, West Germany, and Italy, to construct gas pipelines to their countries. Through "gas for pipes" deals, western gas companies paid for the construction of pipelines in return for imports of Soviet gas. These mutually beneficial agreements provided the Soviets with the infrastructure needed to access new markets and western European countries with a guaranteed source of energy. From the late 1960s through the 1980s, these East-West partnerships focused on constructing a pipeline network to transport gas from the significant reserves in western Siberia to

---

8. For a detailed history of the development of the Soviet pipeline network, see Högselius 2013.

European customers. By end of the Cold War, three major pipelines transported Russian gas to western Europe, making the Soviet Union a key exporter of gas to the region. Due to these infrastructural developments, Soviet gas exports to western Europe more than doubled between 1983 and 1990.[9]

Given this existing trade network, post-Soviet Russia emerged as a significant power in the European gas markets. The former Soviet gas ministry transitioned into a new Russian state-owned gas company, Gazprom, which maintained its partnerships with European gas companies and upheld existing arrangements to supply gas to the West. Additionally, Russia controlled the vast majority of gas reserves in the former Soviet Union, which were mainly located in western Siberia. Other former Soviet republics with significant gas reserves—Azerbaijan, Kazakhstan, Turkmenistan, and Uzbekistan—transported their gas exports through Russia. In addition to its exports to western European countries, Gazprom was the predominant supplier of gas to many former Warsaw Pact allies and other newly independent former Soviet republics.

Sources of Market Power

Given the structure of the European gas market, post-Soviet Russia enjoyed significant market power. First of all, as we have discussed, the use of pipelines to transport gas benefited Russia. The legacy Soviet trunk pipelines allowed gas to move easily from Russian fields to western Europe. This also meant that pipeline networks in eastern Europe and in former Soviet republics were largely designed to import gas from Russia. Moreover, developments in the European pipeline network in the early part of the post-Soviet era tended to largely benefit Russia. Partnerships between Gazprom and western European gas companies continued to move forward with the construction of pipelines between Russia and western Europe. Most notably, Russia began transporting gas from western Siberia to Germany through Belarus and Poland in the late 1990s.[10] Not only did this provide a new route to supply Russian gas to Germany, but it also allowed Gazprom to further penetrate the gas market in Poland.

Even if they so desire, many European countries are limited in their options for importing pipeline gas from non-Russian sources. The construction of new pipelines requires a significant investment of time and money. It takes several years of planning, bureaucratic approval, and construction before a new pipeline can come online. In the coming decades, available gas reserves in the European

---

9. Högselius 2013, 200.

10. Ibid., 213.

Union will continue to decline, so EU customers will likely need to look externally for their gas needs.[11] Additionally, given the practicalities of transporting gas, it is often not economical to build pipelines over extremely long distances. To import pipeline gas, EU countries are largely geographically constrained to suppliers in North Africa, the Eastern Mediterranean, or the Caspian region. Thus, importing countries have no choice but to use the existing pipeline networks in the short term, and they have limited options when searching for alternative sources of pipeline gas in the medium to long term.

In addition to the structure of the pipeline network, Russia has also benefited from the pricing structure in the European gas market. Most of Russia's gas exports to Europe have been sold on long-term contracts that last between ten and thirty-five years.[12] Given the lack of a natural market price for gas, these contracts have typically benchmarked the price of gas to the price of oil. They have also included "take or pay" obligations in which the buyer commits to purchasing a minimum volume of gas.[13] Historically, long-term contracts have facilitated the development of cross-border gas trade between national monopolist gas companies. For gas exporters, these contracts provide a guaranteed flow of income that makes it cost-effective to invest in the significant infrastructure needed to transport gas. For gas importers, they provide a long-term guaranteed source of energy. Thus, long-term contracts provide a level of certainty to the trade of a critically important resource.

This contract system gives significant market power to sellers. First of all, it limits the potential for competition among gas suppliers that one would likely see in a true market. If a gas importer cannot easily turn to alternative suppliers and needs to reach an agreement in order to keep the gas flowing to its customers, a gas exporter can act similarly to a monopolist. When a gas supplier is able to capture a market in this way, it has the ability to set prices above what they would be in a competitive market. Second, the contract system gives sellers the ability to engage in price discrimination and set different gas prices in different countries. Since each contract is negotiated separately, a gas exporter has the ability to agree to a different price for each of its buyers. While some of this differential in prices may be due to purely economic factors like the relative cost of transporting gas to different locations or the level of demand in a given market, differential price setting also opens the door for rent-seeking behavior. Additionally, exporters have the ability to link gas prices to other political or economic issues by using the

---

11. Grigas 2017, 142.

12. Dickel et al. 2014.

13. Vavilov and Trofimov 2015a, 161.

promise of low prices as a reward or the threat of high prices as a punishment. In this way, price setting can be used as a coercive tool in international relations.

Gazprom's ability to set prices in a given country depends upon the market structure. In western European countries less dependent upon Russian gas, Gazprom has less of an ability to set prices. For its gas trade with these countries, Gazprom has traditionally relied upon long-term contracts in which the price of gas was benchmarked to the price of oil. These contracts provide incentives for consumers to choose gas over oil and give suppliers some price-setting power, but prices are also affected by fluctuations in the global oil market.[14] In some cases, Gazprom has been able to adjust the formulas in these contracts on an ad hoc basis, but it is more constrained by competitive pressures when negotiating prices with countries that have access to alternative suppliers. On the other hand, in eastern European countries in which Gazprom is largely a monopoly supplier, Gazprom has enjoyed greater price-setting ability. Until the middle of the first decade of the 2000s, Gazprom did not even benchmark gas prices in former Soviet republics to the price of oil. Instead, it renegotiated the price of gas with these countries on an annual basis.[15] This gave Gazprom significant power to set gas prices in these countries.

Political Benefits of Market Power

Russia has used Gazprom's power to set prices to its political advantage. Most of the natural gas produced by Gazprom each year is used for domestic consumption within Russia. The Russian government has long required Gazprom to sell gas to these domestic customers at artificially low prices. To make up for the shortfalls in the domestic market, Gazprom relies upon revenue from exports to Europe. In the years following the breakup of the Soviet Union, Gazprom typically charged EU customers significantly higher gas prices than customers in former Soviet republics. In exchange for these lower prices, Russia often expected political cooperation. This pricing strategy aimed to both keep these countries under Moscow's umbrella and prevent them from shifting away from reliance on natural gas for their energy needs.[16] Even with these lower prices, many of these countries accumulated significant debt to Gazprom.

During the Putin era, Russia has increased its use of its price-setting ability in energy markets to achieve political goals. In the first decade of the 2000s, in the

---

14. Vavilov and Trofimov 2015b, 113–114.

15. Ibid.

16. Hedlund 2014, 26; Vavilov and Trofimov 2015b, 113–114.

face of increasing oil prices, Gazprom began a process of shifting gas prices in the former Soviet republics to align with European prices. Both Gazprom's credible threat to raise gas prices and the level of debt accumulated by many former Soviet republics gave Gazprom leverage in the annual negotiations of gas contracts. Gazprom could adjust the speed of gas price increases or forgive debt in exchange for political and economic concessions. A central goal of Gazprom in these negotiations was increased control over the gas infrastructure in former Soviet republics, which would augment Russia's power in the European gas market. As we will see in the following section, Gazprom achieved mixed success in this endeavor, but these negotiations highlight the political and economic leverage that accompanies the ability to set prices. Given this, Russia has significant incentives to take advantage of any opportunities to maintain and expand its market power.

Challenges to Russia's Market Power

While Russia largely enjoys a privileged position in the European gas market, challenges to its market power have emerged in past two decades. Several actual and potential shifts in the market structure threaten Russia's ability to set prices for political advantage or to seek rents. First, transit states like Ukraine and Belarus have the potential to disrupt the flow of gas from Russia to European customers. Most notably, this threat manifested itself in a pair of gas disputes between Russia and Ukraine in the winters of 2006 and 2009 that led to temporary gas shutoffs in parts of Europe. Second, additional competitive suppliers have emerged. These include other former Soviet republics with significant gas reserves and producers of LNG. Third, the European gas market has begun to see a shift from long-term gas contracts to something closer to a true market in which gas is traded at hubs on spot prices. This change in the pricing structure, along with efforts by the European Union to liberalize gas markets, creates the potential for increased export competition between gas suppliers, which would reduce Gazprom's price-setting capability.

Transit Countries

When the major gas pipelines were constructed in the Soviet era, they all traversed through the Ukrainian Soviet Socialist Republic to Czechoslovakia.[17] Using this common entry point into Europe provided an efficient way to

---

17. The only exceptions were smaller pipelines to Finland, Greece, and Turkey (Högselius 2013, 212).

construct a pipeline network to move Soviet gas to European customers. The location of this supply route did not matter when Ukraine was part of the Soviet Union. However, after the breakup of the Soviet Union, this meant that practically all Russian gas exports to Europe would have to transit through a newly independent Ukraine. Until 1999, when the Yamal-Europe pipeline connecting Russia to Germany through Belarus and Poland was completed, Gazprom had no alternative routes to supply gas to most of its western European customers other than through Ukraine. Even after the Yamal-Europe pipeline reached full capacity, 80 percent of Russia's gas exports to western Europe continued to move through Ukraine.[18]

Ukraine's position as a key transit state in the European gas market provided it with bargaining power. The need to negotiate the level of transit fees paid by gas suppliers along with the price of domestic gas imports provides transit countries with opportunities to potentially extract rents from supplier countries. Additionally, the ability to potentially disrupt gas flows gives these transit countries significant leverage. Since any uncertainty created by the threat of potential disruptions in the gas supply can frighten gas importers and cause them to pursue alternative energy sources, suppliers may need to take steps to ensure the continued cooperation of transit countries. Thus, the presence of transit countries can threaten the market power of supplier countries, along with their ability to extract rents.

The "gas wars" that erupted between Russia and Ukraine in 2006 and 2009 illustrate the issues that can arise in the relations between supplier and transit countries. In the years following independence, Russia and Ukraine had been able to reach mutually acceptable gas trade agreements. Gazprom would provide transit fees to Ukraine, usually in the form of gas, and also sell gas to Ukrainian customers at lower prices than those charged in western European markets. In the first decade of the 2000s, Gazprom began "Europeanizing" gas prices in former Soviet Republics, including Ukraine. At the same time, Gazprom moved to end the barter process and instead require cash payments for both transit fees and gas purchases between the countries. Under the barter system, Ukraine had not been able to cover its payments for Russian gas, and by 2004, the Ukrainian gas company Naftogaz had accumulated a $1.7 billion debt to Gazprom. On top of these disputes over prices and debt repayment, Gazprom was also concerned that Ukraine was syphoning off Russian gas and illegally selling it to Europe.[19]

18. Abdelal 2013, 430.

19. Vavilov and Trofimov 2015b, 123.

In 2005, Ukraine was paying an effective price of about $77 per thousand cubic meters (tcm) for its gas imports from Russia, compared to the price of about $260 per tcm in the European Union.[20] When Gazprom pushed for Ukrainian customers to pay higher prices that were more in line with those in Europe, Naftogaz balked. Negotiations between the two sides failed to reach an agreement on the terms of the annual gas contract by the year's end, and Gazprom decided to cut off gas exports to Ukraine on January 1, 2006. Since Gazprom still needed to fulfill its contracts to supply gas to other European customers, in practice this meant that it reduced the flow of Russian gas in the pipelines by the amount that would be provided to Ukrainian customers. However, Ukraine continued to syphon gas from the pipeline to meet its domestic needs, reducing the supply of gas available to European countries in the middle of winter. The standoff lasted only lasted three days, and Gazprom and Naftogaz were able reach an agreement on transit fees and gas prices in Ukraine on January 4. In the end, Naftogaz agreed to purchase gas from Russia at an effective price of $95 per tcm, which would increase to $130 per tcm by the following year.[21]

Three years later, a similar gas dispute broke out between Russia and Ukraine, but this one had more serious consequences for European consumers. In the intervening years, Gazprom had continued to push for Ukraine to pay gas prices on par with those paid in Europe. However, Ukraine was not able to keep up with its payments at the existing gas prices. By late 2008, Ukraine had accumulated a debt totaling $2.4 billion. It paid $1 billion of this debt in December 2008, and Russia demanded that it repay the remainder. As with the earlier dispute, the two sides were unable to reach an agreement on a contract for the new year. In response, on January 1, 2009, Gazprom reduced gas exports to Ukraine by 110 million cubic meters per day, which reflected the amount meant for Ukrainian consumption. Ukraine continued to divert gas for its own consumption, which led to shortages in Hungary, Romania, and Poland. On January 7, Russia escalated the dispute and cut off all gas supplies into Ukraine. On January 19, negotiations between the Russian and Ukrainian leaders finally led to an agreement on prices and transit, in which Ukraine would pay an effective price of $230–$250 per tcm, and natural gas once again began flowing from Russia to Europe through Ukraine.[22]

The gas disputes between Russia and Ukraine had significant political and economic effects on the European gas market.[23] First, both Russian suppliers and

---

20. Belyi 2015, 148.

21. Grigas 2017, 181.

22. Ibid., 183.

23. Abdelal 2013.

European consumers became aware of the issues of depending upon transit countries like Ukraine to transport gas from Russia to Europe. Gazprom was not able to directly control the flow of gas from the source to its contracted customers. The potential for Ukraine to disrupt the transportation of gas and the need to negotiate transit issues inhibited Russia's ability to capitalize on its market power. As we will discuss further, the 2006 gas crisis led to an increased push by Gazprom and some eastern European gas companies to develop pipeline infrastructure that could bypass Ukraine and transport gas directly from Russia to EU customers.

Second, the gas crises highlighted the vulnerability of countries highly dependent upon Russian gas. The disruptions in the supply of gas hurt Gazprom's reputation. In the eyes of some European leaders, Gazprom was now seen as an unreliable supplier of natural gas. From their perspective, too much dependence on Russia created a significant threat to the energy security of the European Union. In response, some European countries began to look for alternative sources of energy that would reduce their dependence on Russian gas and thus their vulnerability to disruptions of gas deliveries caused by Russian political machinations or instability along the Ukrainian supply line. We now turn to some of these potential competitors to Russian pipeline gas.

Competition from Pipeline Gas Suppliers

As we have discussed, many central and eastern European countries are largely dependent upon Russian gas for historical and geographical reasons. In the past half century, Soviet/Russian and European gas companies have constructed major pipelines to transport gas from Russian sources to European consumers. Construction of new pipelines requires a significant time and capital investment, and technology limits the feasible distance that a pipeline can run, so the location of these existing fixed pipelines limits the options that countries have to import gas. However, potential competitors to Russian pipeline gas have emerged. First, European companies pursued projects to construct pipelines to import gas from non-Russian suppliers, most notably Azerbaijan. Second, European countries have increased their capacity to import LNG. Finally, the advent of fracking technology and the shale gas boom have made available new sources of natural gas worldwide that could be transported to Europe as LNG.

In addition to creating transit countries in the supply chain from Russia to European customers, the breakup of the Soviet Union also led to the emergence of new potential export competitors for Russia. While the majority of Soviet gas reserves were located in Russia, other former Soviet republics also had significant natural gas reserves. These Eurasian countries—Azerbaijan, Kazakhstan, Turkmenistan, and Uzbekistan—now had the potential to become rival suppliers of gas. In the years following independence, the gas industries in these countries did not provide significant challenges to Russia, as the gas pipelines from these countries connected to pipelines in Russia. Gazprom was the primary buyer of Eurasian gas, and any gas that might be sent to non-Russian customers from these countries had to pass through Russia. Due to this monopsony power and transit monopoly, Gazprom had significant price-setting ability in the gas trade with Eurasian countries.[24]

However, Russia's position vis-à-vis these export countries began to decline somewhat, beginning in the first decade of the 2000s, when they began to pursue opportunities to sell gas directly to non-Russian consumers. Among these Eurasian gas producers, Azerbaijan has taken the most significant steps toward supplying gas to the European Union. To date, Azerbaijan primarily competes with Russia in the Black Sea gas market. The South Caucasus pipeline (also known as the Baku-Tblisi-Erzurum pipeline) runs from Azerbaijan through Georgia to eastern Turkey. It went online in 2006 and opened up the Turkish market to Azerbaijani gas. By 2017, Azerbaijan supplied about 12 percent of gas imports into Turkey, behind Russia and Iran.[25] Additionally, in 2006, Georgia decreased its imports of Russian gas and instead turned to its neighbor, Azerbaijan, as its primary gas supplier.[26] By 2015, Georgia depended upon Azerbaijan for 90 percent of its gas imports, compared to only about 3 percent from Russia.[27]

Beyond being an access point to the Georgian and Turkish markets, the South Caucasus pipeline is also the first stage of the planned Southern Gas Corridor, which will transport gas from the Shah Deniz gas fields in Azerbaijan to Europe through a series of pipelines that bypass Russian territory. The Trans-Anatolian Natural Gas Pipeline (TANAP), which was completed in 2018, connects the South Caucasus pipeline to the border between Turkey and Greece. Initially, consortia of gas companies put forward two competing pipeline projects to transport Azerbaijani gas from Turkey to EU markets. The Nabucco pipeline was to run from Turkey through Bulgaria, Romania, Hungary, and Austria and was proposed by a consortium of gas companies from these countries. This pipeline would have largely served countries relatively dependent on Russia gas. A separate partnership of western European gas companies proposed the Trans-Adriatic Pipeline (TAP) that would run through Greece and then across the Adriatic Sea

---

24. Nygren 2008.

25. BP 2018. Russia and Iran accounted for 51.4 percent and 16.6 percent of these imports, respectively.

26. Grigas 2017, 212; Vavilov and Trofimov 2015b, 116.

27. Grigas 2017, 212.

to southern Italy. In June 2013, the Shah Deniz company decided in favor of TAP, and the Nabucco project was scrapped. Construction of TAP is expected to be completed in late 2020.[28] Once this final leg of the Southern Gas Corridor is online, Azerbaijani gas will begin flowing into EU markets.

Like Azerbaijan, Central Asian gas producers have also pursued additional markets outside of Russia. As the number of potential buyers of gas from Central Asian countries has increased, Russian monopsony power in this region significantly diminished. In 2005, Turkmenistan negotiated a contract to export gas to Ukraine. Gazprom was able to block this deal, but at the cost of agreeing to pay a higher price in hard currency for Turkmenistan's gas.[29] During this time period, Turkmenistan also pursued options to enter the Chinese and European markets. In 2006, Turkmenistan and China agreed to build a gas pipeline between their countries. Construction of the Central Asia–China pipeline began in 2007, and it went online in 2009.[30] In light of the 2006 Russian-Ukrainian gas dispute, European actors also revived interest in building a Trans-Caspian pipeline that would connect Turkmenistan with Azerbaijan and allow for the possibility of transporting Central Asian gas to Europe. However, this project has been blocked by both Russia and Iran and still remains on the drawing board.

During the first decade of the 2000s, as these rival markets were coming into play, Russia was relying on Central Asian gas to fulfill its commitments to satisfy an increasing demand for gas in Europe.[31] However, when the 2008 financial crisis led to a decrease in the demand for gas, Gazprom found paying a high price for Turkmenistan's gas unprofitable and began reducing its imports. In response, Turkmenistan—and to a lesser extent, Uzbekistan—turned its focus to the Chinese market.[32] Today, in contrast to the 1990s and early 2000s, Russia's status as a buyer of Central Asian gas is greatly reduced. Russia still remains the largest importer of gas from Kazakhstan, and it receives over one-third of the gas exported from Uzbekistan. However, the largest Central Asian gas producer, Turkmenistan, has ceased shipping gas to Russia and now primarily exports to China.[33] Rather than providing increased competition in Europe, Central Asian producers will likely serve as future rivals for Russia in the Chinese gas market,

28. Ibid., 160–161.

29. Vavilov and Trofimov 2015b, 124.

30. Hedlund 2014, 78.

31. Nygren 2008, 10.

32. Grigas 2017.

33. BP 2019.

as Gazprom began supplying gas to China through the Power of Siberia pipeline in 2019.

Competition from LNG Suppliers

In additional to alternative suppliers of pipeline gas, Russia also faces increased competition from producers of LNG. As its name implies, LNG is natural gas that is held in its liquid state at very low temperatures. Because LNG takes up much less volume than natural gas in a gaseous state, it is economical to store it in large quantities and ship it via tankers.[34] Compared to pipeline gas, potential suppliers of LNG are less constrained by geography. Since LNG can be transported by tankers, it can be shipped from any country in the world with access to a port. The largest suppliers of LNG to Europe are Qatar, Algeria, and Nigeria.[35] As European countries expand their capability to import LNG, they increase the number of potential suppliers of gas to Europe and, in turn, the number of potential export competitors that Russia faces.

While LNG constitutes a potential alternative to Russian pipeline gas for European consumers, it has not yet significantly dented Russia's market share. Currently, LNG makes up about 22 percent of gas imports into European countries, and there are challenges to the increased use of LNG in Europe.[36] First, LNG is generally more expensive than pipeline gas. Thus, customers would need to pay a premium if they were to abandon pipeline gas for LNG. Second, importing LNG requires significant investment in infrastructure. To transfer LNG from tankers to pipelines (and on to customers), regasification terminals at ports are needed to convert LNG from its liquid state to its gaseous state. Construction of the regasification terminals requires significant time and capital, so countries must be willing to devote resources to building this infrastructure before they can import LNG for their energy needs. Thus, countries cannot immediately shift from pipeline gas to LNG.

To date, most European LNG terminals are located in southern and western Europe, while there is less capacity for northern and eastern European countries to import LNG. In recent years, some countries that are more highly dependent upon Russian gas have moved to build LNG terminals to gain access to alternative suppliers. Lithuania opened its first LNG terminal in 2014 at the port of Klaipėda, which for the first time allowed this Baltic country to import gas from

---

34. Grigas 2017, 42.

35. BP 2019.

36. Ibid.

suppliers other than Russia. Neighboring Poland began importing LNG at its own terminal at Świnoujście in 2016.[37] Thus, the emergence of alternative sources of gas from LNG suppliers provides a significant challenge to Russia's market power. Recognizing this, Russia has ramped up its own production of LNG and has targeted new export markets. In 2018, Russia began exporting LNG to several European countries, including the United Kingdom, France, Spain, and Belgium. As a result, Russian LNG exports increased 61.5 percent from the year before.[38] However, LNG still makes up a small percentage of Russia's gas exports.

In addition to the growth of LNG production, the advent of hydraulic fracturing ("fracking") technology to profitably extract gas from shale gas fields has had a sizable effect on energy markets. While the shale gas boom has largely been focused in the United States, it has had ripple effects around the globe. Due to increased use of fracking, the United States is now the largest natural gas producer in the world. As the shale boom in the United States has led to an increased supply of natural gas globally, other producers of LNG have needed to find alternative export markets. Many of them turned to Europe, which increased competition for Russia. In the long term, if it were to increase its capacity to export LNG, the United States itself could become a rival supplier of gas in Europe. At this point, the United States does not export a significant amount of gas to Europe, and most of these exports are sent to southern Europe and do not generally compete with Russia. However, in 2017, the United States began shipping LNG to Poland and Lithuania, two countries that rely upon Russian pipeline gas for their energy needs.[39] Thus, the US may emerge as a future export competitor to Russia in the European gas market.

Changes in the European Gas Market

A final set of challenges to Russia's market power comes from within the European Union. In particular, an ongoing shift in the pricing structure for natural gas toward hub pricing and greater EU regulatory activity in energy markets raises competitive pressures on Russia as a gas exporter to the region. As we discussed earlier, the price of natural gas in Europe has historically been set by long-term contracts that are typically pegged to the price of oil. These contracts between gas producers and consumers help to reduce uncertainty about the supply of energy and make it profitable for market actors to invest in the infrastructure needed

---

37. King & Spalding LLP 2018.

38. BP 2019.

39. Stytus 2017.

to transport gas. Most importantly for Russia, the use of these contracts reduces competition between suppliers and provides opportunities for price-setting capabilities. However, in recent years, the European gas market has begun to shift to hub pricing based upon the supply and demand of natural gas. This shift threatens Gazprom's ability to set prices and could reduce opportunities for Russia to extract rents through the gas trade.

Hub pricing for natural gas was initially developed in the United States. In this model, the price of natural gas is based upon the actual supply and demand of gas at hubs in the pipeline network. These can be either virtual hubs where gas can be traded virtually or a physical hub where multiple pipelines meet. At a hub, it is possible to engage in spot trade, where gas is sold for immediate delivery at the current market price, rather than a price set in a long-term contract. Thus, the seller has much less ability to set the price paid for gas. The United Kingdom established the first gas hub in Europe in 1996. From there, the practice made its way to the continent, most notably in the Netherlands and Belgium. Currently, the most mature gas hubs are in northwestern Europe, while hub pricing is much less developed in other parts of the continent, especially in the central and eastern European countries that are highly dependent upon Russian gas.[40] Gazprom has been reluctant to abandon both its use of long-term contracts and oil-linked pricing. However, in recent years, competitive pressures from lower hub prices have at times led Gazprom to provide price discounts to its contracted consumers and to institute limited spot pricing.[41] While there is not yet a single gas market in Europe, this move toward market-based pricing increases competition between gas suppliers on the continent and threatens Russia's ability to extract rents.

In addition to these shifts in the gas pricing structure, the European Union has taken steps to increase the competitiveness of the gas market in Europe. The goal of these regulatory measures is to move the continent closer to a single EU energy market. The most important of these measures is the Third Energy Package, which was enacted by the European Commission in 2009. This legislation included three main provisions that affected Gazprom's operations within the European Union: ownership unbundling, third-party access, and "third country" certification requirements.[42] Ownership unbundling prohibits a single company from controlling both the production of energy resources and the transmission system for those resources. Thus, the same company cannot be both the supplier of natural gas and the owner of the pipelines used to transport that

---

40. Heather 2015; Heather and Petrovich 2017.

41. Mitrova 2015; Pinchuk, Astakhova, and Vukmanovic 2015.

42. Grigas 2017, 154–156.

gas. Third-party access regulations require that a company's pipelines are accessible for use by competing gas companies. This limits Gazprom's capacity to transport gas through the pipelines in Europe that it partially owns and operates. The third-country provision—also known as the "Gazprom clause"—requires that the European Union certify that gas transmission owners or operators from non-EU countries comply with these regulations and "will not put at risk the security of energy supply of the Member State and the Community."[43] While these regulations have been unevenly applied, they have hampered Gazprom's ability to expand its operations within the European Union, and the restrictions on Gazprom's ability to control the transportation of Russian gas to European customers reduce its price-setting capability.

The threat of potential disruptions by transit countries, the emergence of export competitors, and the shift of the pricing structure away from long-term contracts all potentially threaten Russia's position in the European gas market and its ability to seek rents by setting prices. Given the cost and infrastructure requirements of switching to alternative sources of pipeline gas or LNG, European countries currently dependent upon Russian gas are likely to remain so in the short term. Additionally, European demand for natural gas is rising, and current projections indicate that the production of gas by the European Union will decline in the coming decades.[44] Thus, Russia has the potential to remain in a privileged market position going forward. However, given the desire of many European states to reduce their dependence upon Russian gas and the emerging shifts in the gas market we have outlined, Russia has needed to take active steps to preserve its market power in the medium to long term. For the remainder of this chapter, we turn our attention to the various strategies that Russia has pursued in its attempts to achieve this goal.

Maintaining Market Power and Pipeline Politics

To maximize its capability to set prices and engage in rent-seeking behavior, Russia has a significant motivation to maintain and potentially enhance its position in the European gas market. Ideally, Russia would like Gazprom to come as close as possible to monopolizing both the supply and transit of natural gas to Europe. While this extreme goal is not reasonably possible, Russia continually formulates strategies to maximize its share of the market. These strategies largely fall within three categories. First, Russia takes steps to control as much of

---

43. European Union 2009, 107.

44. Grigas 2017, 142.

the supply chain as possible through vertical integration and the construction of pipelines that bypass transit countries. Second, Russia attempts to block or control potential rival suppliers of gas to Europe. Finally, Russia seeks to expand the size of the natural gas reserves under its control. In this section, we consider the first prong of Russia's strategy portfolio, which is commonly referred to as "pipeline politics." In the following section, we then turn our attention to the other two prongs in the context of property rights disputes with Russia's neighbors.

Vertical Integration

The most basic way for Russia to maximize its power in the European gas market would be for Gazprom to control as much of the supply chain as possible. If Gazprom could control both the production and transport of gas, it would be in an even better bargaining position when negotiating gas contracts. With such market power, Gazprom would have an increased ability to set prices. Additionally, greater control of the supply chain would reduce the ability of transit states to extract rents and reduce Gazprom's profit margin. In the early post-Soviet era, Gazprom pursued such a vertical integration strategy and had mixed results. Within the European Union, Gazprom was unable to gain control of downstream pipes. However, it eventually had some success with these efforts in other countries in the former Soviet Union.

After the breakup of the Soviet Union, Russia was the primary supplier of gas to many of the newly independent republics. Initially, gas prices in these countries were significantly lower than those in western Europe. In the 2000s, Gazprom moved to reduce these subsidies and raise gas prices in the former Soviet Union to be more in line with those it charged in western Europe. As part of these negotiations, it also attempted to gain greater control of downstream pipelines in these countries, many of whom had racked up significant debt to Gazprom since independence. In many cases, Gazprom offered to forgive this debt and/or delay gas price increases in exchange for ownership of the gas pipelines in these countries. Through such negotiation tactics, Gazprom was able to gain control of pipelines in the countries of Moldova, Belarus, and Armenia. However, similar efforts were largely unsuccessful in Ukraine and Georgia.[45]

In 1998, Gazprom made its first successful attempt to control downstream pipelines when it gained control of Moldova's pipe transmission network. A major pipeline supplying gas to southeastern Europe traverses both Ukraine

---

45. In addition to these cases, Gazprom temporarily gained a minority stake in the pipelines in Estonia, Latvia, and Lithuania.

and Moldova. As part of negotiations over Moldova's gas debt, Gazprom emerged with a majority share (50 percent plus one) in the Moldovan gas company, Moldovagaz. As part of later negotiations over gas prices, Gazprom gained control of an additional 13.44 percent stake in Moldovagaz in 2006 in exchange for an agreement to delay increasing Moldova's gas prices to European levels until 2011.[46] In this way, Gazprom was able to use the leverage that it had through its price-setting ability to gain control of the segment of the gas pipeline running through Moldova. However, without control of the portion of the pipeline running through Ukraine, Gazprom still had to contend with negotiating with a major transit country along this route. Thus, the Moldovan deal can be seen as a first step in a larger vertical integration strategy.

However, Gazprom was unable to translate its success in Moldova to its dealings with Ukraine. In comparison to Moldova, Ukraine held significantly more bargaining power due to its role as the primary transit state for Russian gas. In the decade after the breakup of the Soviet Union, almost all Russian gas exports went through Ukraine, and even when the Yamal-Europe pipeline through Belarus reached full capacity in the first decade of the 2000s, 80 percent of Russia's gas exports to Europe continued to transit through Ukraine.[47] Thus, Gazprom has historically been highly dependent upon Ukraine to be able to reach its European customers. In the 1990s, Gazprom attempted to gain control of the Ukrainian gas company, Ukrtransgaz, which controls Ukraine's gas infrastructure, in exchange for forgiving Ukrainian gas debt. When these entreaties failed, Gazprom continued to pursue various strategies to gain control of the pipelines running through Ukraine.

Beginning in 2002, as part of negotiations over gas prices and debt payment, Russia regularly offered up the idea of creating an international gas consortium, which would include Gazprom, to take control of gas transit through Ukraine. Ukraine repeatedly rejected these offers. As we discussed earlier, these negotiations between Russia and Ukraine broke down on multiple occasions, leading to the "gas wars" of 2006 and 2009. Even after these crises were settled, Gazprom continued to offer proposals to gain control of Ukraine's pipelines, but Ukraine was never willing to make such a concession.[48] Unlike Moldova, Ukraine was in a better bargaining position to resist Russia's coercive tactics and hold on to its gas infrastructure. Given that Gazprom was unable to vertically integrate gas transit

---

46. Bruce 2007.

47. Abdelal 2013, 430.

48. Vavilov and Trofimov 2015b, 122–127.

through Ukraine into its own operations, Russia has instead had to turn to alternative strategies to mitigate its dependence upon Ukraine as a transit state.

Russia's vertical integration strategy was much more successful in its other major transit state, Belarus. Before the construction of the Nord Stream pipeline, about 20 percent of Russia's gas exports transited through Belarus.[49] As with Moldova and Ukraine, in the early 2000s, Russia attempted to negotiate control over the gas infrastructure in Belarus in exchange for low gas prices and/or a reduction in Belarus's gas debt. When these negotiations reached a deadlock in February 2004, a short gas dispute resulted in which Russia cut off gas flows to Belarus, and Belarus in turn shut down the transit of Russian gas to Poland and Lithuania.[50] The dispute highlighted Belarus's potential leverage as a transit state, and at this point Gazprom was unsuccessful at increasing its control over the flow of gas through Belarus. In the wake of another potential dispute at the end of 2006, Gazprom was able to purchase a 50 percent stake in the Belarussian gas transport company, Beltransgaz, for $2.5 billion, part of which was paid through forgiveness of Belarus's debt. However, at this point Gazprom did not have a controlling share in the company.[51]

By 2011, however, Belarus found itself in a weaker bargaining position. For one, the Nord Stream pipeline that could directly transport gas to Germany went online, which decreased Belarus's leverage as a transit state. More importantly, Belarus was suffering from a severe financial crisis and was unable to obtain loans from the International Monetary Fund due to election irregularities the previous year. Russia provided credit to Belarus to prevent it from falling into bankruptcy. As part of this deal, Russia pressured Belarus to privatize many of its energy assets, and Gazprom purchased the remaining 50 percent of Beltransgaz.[52] At this point, the Russian company gained complete control over the transit of gas through Belarus.

While it is less critical than Ukraine and Belarus, Georgia is a transit state for Russian gas heading to Armenia. In 2006, Gazprom attempted to use its price-setting leverage to gain control of gas flows in both Georgia and Armenia. Control of these pipelines would allow Gazprom to connect to pipelines between Iran and Russia and would hinder efforts to build alternative pipelines to Europe that would bypass Russian territory.[53] In both Armenia and Georgia, Gazprom threatened to increase gas prices to European levels if these countries did not

49. Ibid., 118.

50. Ibid., 119.

51. Grigas 2017, 198; Vavilov and Trofimov 2015b, 120.

52. Grigas 2017, 199; Vavilov and Trofimov 2015b, 121–122.

53. Smith Stegen 2011, 6509.

cede control of their gas infrastructures. Armenia was dependent upon Russia for its gas supply and had no viable outside options. The only pipelines going into Armenia originated in Russia. Moreover, Armenia does not have diplomatic relations with its gas-producing neighbor, Azerbaijan, due to a long-standing dispute over control of the Nagorno-Karabakh region.[54] Thus, it could not credibly turn to Azerbaijan as an alternative source of natural gas, and a pipeline to import gas from Iran had not been constructed. With such little bargaining power, Armenia was willing to sell Gazprom a controlling stake in its national gas company, ArmRosGazprom, in exchange for a two-year freeze in gas prices.[55]

The situation was different in Georgia. Russian-Georgian relations had been strained since the Rose Revolution in 2003. So, when Russia demanded that Georgia either cede control over its gas pipelines to Gazprom or face a significant gas price increase to European levels, Georgia was less amenable to making any concessions. While Georgia's leverage as a transit state was likely not too great, given that it was the intermediary to a small market in Armenia, it had an outside option not available to its neighbor to the south. The South Caucasus pipeline from Azerbaijan to Turkey through Georgia went online in 2006. This opened up an alternative source of natural gas for Georgia. Azerbaijan was willing to sell gas to Georgia for a lower price than that demanded by Gazprom. Thus, rather than ceding control of its pipelines to Gazprom or agreeing to pay European-level prices for Russian gas, Georgia instead decided to stop importing gas from Russia and instead buy its gas from Azerbaijan.[56] The presence of an export competitor reduced Russia's price-setting abilities and its leverage over Georgia. Given this, not only was Russia not able to vertically integrate its operations in Georgia, it ended up losing it as an export market. Moreover, Georgia became a critical transit state for a rival gas supplier, Azerbaijan.

Bypass Pipelines

In addition to its attempts to gain control of existing pipelines in transit states, Gazprom has also worked to construct new pipelines that bypass these transit states altogether. In partnership with western European gas companies, Gazprom developed plans for pipeline projects that would avoid transit through Ukraine. In the wake of the Russian-Ukrainian gas crises, gas companies in Germany, Italy,

---

54. Grigas 2017, 207.

55. Nygren 2008, 12.

56. Grigas 2017, 212; Nygren 2008, 12.

and France ramped up these efforts to develop more direct pipeline routes.[57] This would streamline the transportation of gas from Russia to western Europe and help insulate western European countries from any disruptions that might arise from any future gas disputes between Russia and Ukraine. Initially, these partnerships focused on two major bypass pipeline projects: the Nord Stream pipeline between Russia and Germany and the South Stream pipeline between Russia and southern Europe.

The Nord Stream pipeline was a multinational joint project between Gazprom, two German gas companies (E.ON and Wintershall), Nederlandse Gasunie, and Gaz de France Suez (now Engie). The initial Nord Stream pipeline, which went online in 2011, transports gas directly from Vyborg, Russia, to Greifsawald, Germany, through the Baltic Sea. In 2017, the pipeline transported 51 billion cubic meters of natural gas to EU customers, accounting for over 30 percent of all Russian pipeline exports to the European Union.[58] As a follow-up, this consortium (along with OMV and Royal Dutch Shell) has moved forward to construct a second pipeline, the Nord Stream 2, which would follow the same route as the initial pipeline.

While these western European gas firms have been eager to build the Nord Stream pipelines, eastern European countries have largely opposed their construction.[59] In particular, Poland fears that the Nord Stream pipelines will reduce Gazprom's reliance on the use of the Yamal-Europe pipeline that transports gas from Russia to Germany through Belarus and Poland. This will reduce Poland's status as a transit state and decrease its market power vis-à-vis Russia. However, efforts by Poland and other similarly concerned eastern EU countries to stymie the Nord Stream 2 project have been unfruitful, and it is expected to be completed in 2021.

The South Stream project was a joint project between Gazprom and the Italian gas company, Eni, that aimed to construct a pipeline along a southern route that would bypass Ukraine. This pipeline was to connect Russia to Bulgaria through the Black Sea and then continue on to Austria. In addition to the desire to bypass Ukraine, Gazprom also viewed this project as a challenge to the Nabucco pipeline project to bring Azerbaijani gas into southeastern Europe from Turkey.[60] The South Stream and Nabucco pipelines aimed to supply gas to many of the same

57. Abdelal 2013.

58. BP 2018; Nord Stream 2018.

59. Harper 2017.

60. Hedlund 2014, 82–83.

countries, including Bulgaria, Hungary, and Austria.[61] This would have made Azerbaijan an export competitor to Russia in a region where many countries are largely dependent upon Russia for their gas imports.

However, the South Stream project failed due to regulatory problems with the European Union related to the Third Energy Package. In particular, the European Union asked each of the countries to renegotiate their contracts with Gazprom to address issues of ownership unbundling and third-party access.[62] These regulatory requirements threatened the profitability of the project for Gazprom. Given these regulatory issues, along with questions of the economic viability of the project due to low gas prices, Gazprom abandoned the South Stream project in December 2014.[63] However, by this time, Azerbaijan had already rejected the construction of the Nabucco pipeline in favor of the Trans-Adriatic pipeline to southern Italy. Thus, while the South Stream pipeline was never constructed, Russia had succeeded in avoiding a new competitor in many of the markets in southeastern Europe that this pipeline would have served.

As an alternative to the South Stream pipeline, Gazprom has constructed the TurkStream gas pipeline that connects to Bulgaria via Turkey. This project consists of two pipes—one to supply gas to Turkey and the other to export to European markets. In October 2016, Russian President Vladimir Putin and Turkish President Recep Tayyip Erdogan signed a deal to move forward with the project.[64] Gazprom completed construction and brought the pipeline online in January 2020.[65] With the TurkStream pipeline, Russia can now export gas to Bulgaria and other states in southeastern Europe without having to use a route through Ukraine. This project also opens up additional capacity to transport gas to Turkey, which is Russia's second largest export market, and provides direct competition with Azerbaijan's Southern Gas Corridor project.

European Response

In many ways, our focus on Russia's strategy to maximize its market share only considers half of the story. While Russia aims to corner the European gas market, EU countries have an incentive to prevent Russia (or any external supplier) from

---

61. Grigas 2017, 161.

62. Ibid., 157.

63. Reed and Arsu 2015.

64. MacFarquhar 2016.

65. Astakhova and Selzer 2020.

gaining such market power. The Russian-Ukrainian gas crises that led to short-term shutoffs of gas flows to Europe in 2006 and 2009 highlighted the risks of European dependence on Russian gas imports. To this end, many European actors have attempted to counter Russian efforts to consolidate its market share. Over the past decade or so, EU-Russian relations have followed a pattern of Russia taking steps to corner the gas market and European countries attempting to find ways to reduce their long-term dependence on Russian gas. These responses to Russia's pursuit of market power include both policymaking at the EU level and investment in new infrastructure and technology.

The European Union has taken several steps that potentially reduce the continent's dependence on Russian gas. First, the European Union has aimed to reduce the overall consumption of natural gas on the continent. In 2008, the European Parliament passed a significant climate and energy package. Known as the "20-20-20" package, this legislation set goals to reduce greenhouse gas emissions by 20 percent, produce 20 percent of energy from renewable sources, and increase energy efficiency by 20 percent by the year 2020.[66] Reaching these goals would reduce the need to import gas from Russia. Second, the European Union has moved to increase its regulation of Gazprom's activities. In 2009, the European Commission adopted the Third Energy Package, which unbundled the ownership of a pipeline and the transiting gas by the same company and required third-party access to pipelines.[67] These provisions hampered the activities of Gazprom and led to the eventual abandonment of the South Stream project. Additionally, in response to a complaint by Lithuania, the European Union instituted an antitrust investigation into Gazprom's activities in central and eastern Europe in 2012. The European Commission reached a settlement with Gazprom in May 2018 that required Gazprom to take measures to ensure the free flow of gas but not pay any fines.[68]

European countries have also invested in infrastructure and technology in an attempt to undercut Russia's market power. Gas companies are actively pursuing projects to build pipelines to import gas to Europe from the Caspian and eastern Mediterranean regions. Additionally, countries such as Poland and Lithuania have built new regasification terminals to gain the capability to import LNG. Within Europe, the European Union has invested in connecting pipelines across borders.[69] It has also invested in reverse-flow technology, which allows gas to flow

66. Grigas 2017, 150.

67. Kandiyoti 2015.

68. Reed and Schreuer 2018.

69. Ridgewell 2018.

in both directions in a pipeline. These infrastructural developments will allow gas to flow much more freely between European countries, which reduces Russia's ability to directly cut off gas to one country. It also opens up the possibility for a more competitive gas market in Europe. For example, in the wake of the annexation of Crimea, Ukraine stopped importing gas directly from Russia but instead used reverse flows to buy Russian gas indirectly from European countries to the west.[70]

We should note that within Europe, governments and firms vary in their willingness to counteract Gazprom's activities in the continent. In general, countries that are more dependent upon Russian gas have pushed harder to undermine Russia's market power, while political and economic actors in countries that have a more symmetric interdependent relationship with Russia have been less willing to block Gazprom at every turn. For example, in the EU negotiations on the Third Energy Package, Poland lobbied heavily for the inclusion of the "Gazprom clause," while Germany and Italy pushed for weakening the clause.[71] Additionally, in the aftermath of the Russia-Ukraine gas crises, gas firms in Germany, Italy, and France pushed forward with new pipeline projects that would bypass Ukraine. These firms have decades-long relationships with Gazprom and view the Russian firm as a trusted business partner. They were more concerned with reducing the threats of gas crises with transit states.[72] On the other hand, eastern European countries more dependent upon Russian gas were more concerned with Russia's ability to use the supply of gas as a coercive tool. They opposed the construction of pipelines that would reinforce their asymmetric interdependent relationship with Russia. In recent years, this intra-European divide has continued as Germany has pushed forward with the Nord Stream 2 pipeline, despite objections from fellow EU members in eastern Europe.

Territorial Expansion in Georgia and Ukraine

In both Georgia and Ukraine, Russian attempts at vertically integrating the natural gas supply chain in the first decade of the twenty-first century largely failed. Given its outside option to turn to neighboring Azerbaijan for its gas needs, Georgia was able to rebuff Gazprom's efforts to gain control of Georgian gas infrastructure in exchange for lower gas prices. In 2007, Georgia largely ceased purchasing gas from Russia in favor of importing gas from Azerbaijan. Ukraine's

---

70. Grigas 2017, 192–193.

71. Brutschin 2015.

72. Abdelal 2013.

position as Russia's most critical transit state and its willingness to escalate gas crises gave it significant bargaining power in negotiations with Russia over natural gas. Thus, Ukraine never found itself in a position in which it needed to cede control of its gas infrastructure to gain an acceptable deal. Moreover, such a concession would greatly weaken Ukraine's bargaining power in future negotiations with Russia, as it would reduce its control over the transit of gas through its territory.

Given its failure to achieve its goals through gas contract negotiations, Russia needed to turn to other strategies in its relations with Georgia and Ukraine to achieve its market power goals. While not all of Russia's tactics on this front have been territorial in nature, a critical component of its market power strategy relies upon ongoing property rights disputes with these neighboring states (see Figure 6.2). Overall, we can generally categorize Russia's strategy in its territorial disputes with Georgia and Ukraine as being one of strategic delay. In the discussion that follows, we first review Russia's market power motivations and territorial aggression in Georgia in the context of the ongoing disputes over the regions of South Ossetia and Abkhazia. We next turn our attention to Russia's motivations and strategy in Ukraine, including the annexation of Crimea and the military intervention in eastern Ukraine. Then, in line with our theory on market power politics, we discuss how institutional and economic constraints shape Russia's decision to pursue a strategy of delay in Georgia and Ukraine.

FIGURE 6.2. Disputed territories in the Black Sea region.

Strategic Delay and Salami Tactics in Georgia

While the Republic of Georgia is not a significant producer of oil or gas, by the early 2000s it had emerged as a key transit corridor for hydrocarbons from the Caspian Sea. Several important pipelines originating in Baku, Azerbaijan, cross through Georgian territory. The South Caucasus (or Baku-Tbilisi-Erzurum) pipeline, which runs from Azerbaijan through Georgia to Turkey, went online in 2006. This pipeline was the initial component of a long-term project to build a series of pipelines that will eventually transport Azerbaijani gas to Europe. Additionally, around this time, regional actors rekindled discussions about a proposed Trans-Caspian pipeline connecting Turkmenistan to the pipeline in Baku, which would allow for the transportation of Turkmen and Kazakh natural gas to Europe.[73] In addition to this gas pipeline network, the Baku-Tbilisi-Ceyhan and Baku-Supsa pipelines transported oil from the Caspian Sea to the Mediterranean and Black Seas, respectively.

Critically for Russia, all of these pipelines bypassed Russian territory and provided potential avenues to supply hydrocarbons from competing suppliers to European markets. Given Georgia's role as a transit corridor of non-Russian hydrocarbons, Russia would benefit from efforts to destabilize its neighbor. By putting Georgia in a state of strategic vulnerability, Russia could raise questions in the eyes of Western actors about the stability of Georgia as a transit state for oil and gas.[74]

"Frozen Conflicts" in South Ossetia and Abkhazia

Russia's territorial strategy to destabilize Georgia largely revolves around its expansionist activities in the breakaway regions of South Ossetia and Abkhazia, both of which have borders with Russia. South Ossetia is a landlocked region located in north-central Georgia, while Abkhazia lies along the Black Sea in the northwestern corner of Georgia. Both regions are home to non-Georgian ethnic groups that have their own distinct languages. During the Soviet era, South Ossetia and Abkhazia were autonomous regions within the Georgian Soviet Socialist Republic. In the aftermath of the breakup of the Soviet Union, separatists in both South Ossetia and Abkhazia pushed for independence from the newly established Republic of Georgia. A sequence of civil wars ensued in South Ossetia (1991–1992) and Abkhazia (1992–1993). Russia actively intervened in

---

73. Hedlund 2014, 123.

74. Allison 2008, 1166.

support of both the Ossetian and Abkhaz rebels and helped mediate settlements to both conflicts.

Since the civil wars of the early 1990s, the situations in South Ossetia and Abkhazia have been described as "frozen conflicts."[75] The disputes surrounding the political and legal status of these regions remain unresolved, as both South Ossetia and Abkhazia have remained largely outside the effective control of the Georgian government since 1993. The Russian-mediated peace settlements allowed the separatists to gain de facto autonomy from Georgia, and Russian troops remained as peacekeepers in both territories. By maintaining these peace-keeping forces in South Ossetia and Abkhazia, Russia helped to perpetuate the uncertain status between these breakaway regions and the Georgian state.[76] However, outside of a brief war in 2008, Russia's activities surrounding these frozen conflicts have largely remained in the gray zone below the threshold of full-scale war. Thus, we can consider Russia's strategy regarding the settlement of the property rights disputes in South Ossetia and Abkhazia as being one of strategic delay.

While strategically delaying the settlement of these conflicts, Russia has engaged in a series of salami tactics to increase its control over both South Ossetia and Abkhazia. In 2000, the Russian government instituted visa requirements for Georgians to work in or visit Russia, but it excluded residents of Abkhazia and South Ossetia from these restrictions. In the following years, Moscow began distributing Russian passports to the South Ossetian and Abkhaz populations, and by 2006, more than 90 percent of residents of the two regions had gained Russian citizenship.[77] During this time period, Russia also increased its military presence in South Ossetia and Abkhazia by shipping military equipment and building military bases. By 2006, the total amount of military arms, ammunition, and equipment in South Ossetia and Abkhazia exceeded that of the Georgian government, and two years later the two regions possessed twice the amount of military equipment as was held by Georgia.[78] Additionally, the Russian govern-ment took steps to ensure that its favored candidates maintained political control of the two breakaway regions.

At the same time, Russia also attempted to use its leverage as Georgia's gas supplier to pressure Georgia to cede its gas transit assets to Russia. In January

---

75. Grant 2017.

76. Nilsson 2018.

77. Illarionov 2009, 61, 63.

78. Ibid., 60.

2006, a series of explosions in North Ossetia damaged the pipelines supplying Russian gas to Georgia, temporarily cutting off Georgia's gas supply in the middle of winter. Georgia claimed that the explosion was a deliberate act of Russian sabotage, while Russian officials blamed the explosion on "terrorists."[79] Russia's use of its "gas weapon" as a coercive tactic when bargaining with other neighbors during this time period, in combination with other circumstantial evidence, makes it likely that this was a deliberate act on Russia's part. In December 2005, the Presidential Administration of Russia summoned the heads of Russian energy companies to see if it was possible to cut off gas supplies to Georgia.[80] Additionally, Georgian President Saakashvili claimed that Russian officials had issued veiled threats in the weeks prior to the explosion, and no terrorist groups claimed responsibility for the explosion.[81] Regardless, the explosion was widely viewed as a Russian pressure tactic, and Georgia refused concede to Russian demands to sell its gas pipelines.[82] As noted earlier, Georgia instead turned to Azerbaijan for its gas needs.

The Russo-Georgian War

Tensions between Russia and Georgia continued to rise, as Russia consolidated its presence in South Ossetia and Abkhazia. On April 16, 2008, Putin signed a presidential decree that established official relations between Russia and the governing administrations in South Ossetia and Abkhazia. A few days later, a Russian military plane shot down an unmanned Georgian drone in Abkhazia. Over the following months, the Russian military increased its mobilization efforts in Abkhazia and South Ossetia. Then, in the first week of August, several armed clashes erupted between South Ossetian separatist militias and the Georgian military.[83] The following week these tensions escalated to a brief international armed conflict between Russia and Georgia.

On August 7, Russian troops entered South Ossetia, and Georgian forces also began attacking the South Ossetian capital of Tskhinvali. The timing of these events is disputed, but regardless, Russian and Georgian troops were soon

---

79. Paton Walsh 2006.

80. Illarionov 2009, 60.

81. Chivers 2006.

82. Smith Stegen 2011.

83. Allison 2008; Popjanevski 2009.

in open combat.[84] Within three days, Russian forces had driven the Georgian military out of South Ossetia. Meanwhile Russian forces and Abkhaz separatists opened a separate front in the war in Abkhazia, and the Russian Black Sea Fleet established a blockade off the Georgian coast. Russian troops then moved farther into Georgian territory, occupying Gori in the central part of the country and the port of Poti and the Georgian military base of Senaki in the west. While on the offensive, Russia decided not to press on to the Georgian capital of Tbilisi. Instead, the two sides agreed to a ceasefire on August 12; the Russo-Georgian war was over in less than a week.[85] Russian troops eventually withdrew from undisputed Georgian territory, but they remained in the breakaway regions.

Integration and Borderization

In the decade since the war ended, Russia has continued to strengthen its ties to South Ossetia and Abkhazia. On August 26, 2008, Russia recognized both South Ossetia and Abkhazia as independent states.[86] Only a small number of other countries followed suit.[87] In 2014, Russia signed a strategic partnership treaty with Abkhazia, and the following year it signed an even more comprehensive integration treaty with South Ossetia. Both of these treaties aimed to integrate the militaries and customs services of these regions with Russia, and they made significant steps toward the political integration of these regions into Russia. However, at this point, Russia has not made attempts to try to formally annex South Ossetia or Abkhazia. By strategically delaying a settlement on the status of these regions, Russia is able to keep its options open and take additional steps at politically opportune moments. For example, proposals to hold a referendum on the integration of South Ossetia into Russia are often floated at times when Georgia is trying to increase its ties with the West.[88]

In addition to these diplomatic agreements, Russia has used its military presence in the disputed regions to expand its territorial control. As part of its agreements with South Ossetian authorities, Russia is responsible for securing the border between South Ossetia and Georgian-controlled territory. In this capacity, Russia routinely engages in a "borderization" strategy that Georgia calls

---

84. Allison 2008; Popjanevski 2009.

85. Allison 2008; Felgenhauer 2009.

86. Allison 2008.

87. Reuters 2018.

88. Nilsson 2018.

a "creeping occupation." As part of the strategy, over the past several years, Russia has physically shifted the boundary posts of the border to expand South Ossetia's territory into Georgia. Most notably, in a land grab in July 2015, South Ossetia gained access to a mile-long stretch of the Baku-Supsa pipeline and pushed the border to within 500 meters of the E60 highway connecting Azerbaijan and the Black Sea.[89] Slowly but surely, Russian encroachments into Georgian territory continue unabated despite objections from Georgia and the West. Given the uncertainty about the actual location of the border and the minor nature of each Russian advance, the Georgian military has largely been unwilling to respond to these incursions with force. Thus, these salami tactics allow Russia to demonstrate its willingness to brazenly advance its interests into its neighbor's territory and highlight the vulnerability of the Georgian state.

Russia has actively pursued a strategy of strategic delay in Georgia. By deploying and maintaining peacekeeping forces in South Ossetia and Abkhazia after the civil wars in the early 1990s, Russia kept these conflicts frozen and the territorial status of the regions uncertain. In keeping these property rights disputes unresolved, Russia has had the opportunity to engage in a series of salami tactics to expand and consolidate its influence and control over South Ossetia and Abkhazia. Notably, outside of the five-day war in 2008, Russia's activities have remained in the gray zone below full-scale armed conflict. Moreover, during the war, Russia deliberately chose not to escalate the conflict and stopped short of attacking Tbilisi. Through these tactics, Russia has been able to put Georgia in a prolonged position of strategic vulnerability, which has hampered Georgia's ability to strengthen ties to the West and potentially makes it a less attractive transit state for natural gas.[90]

Territorial Expansion in Ukraine

As we discussed earlier, Ukraine's role as the primary transit state for Russian gas gave it bargaining power in the negotiations surrounding the gas crises in 2006 and 2009. For this reason, unlike some of its neighbors, Ukraine was unwilling to cede a share of its gas infrastructure to Gazprom. On the other hand, Russia's main source of leverage in these negotiations was its position as Ukraine's primary external supplier of gas. For example, in 2009, all of Ukraine's gas imports came from Russia, and Russian gas constituted more than half of the gas consumed in Ukraine that

89. Witthoeft 2015.

90. Allison 2008.

year.[91] Ukraine had also accumulated a significant debt to Gazprom from its consumption of Russian natural gas, and at the time it had no other available external sources for its gas needs. Under the contract reached between Russia and Ukraine in 2009, Ukraine was paying increasingly higher gas prices, and by 2013, it was paying the highest prices in Europe for Russian gas.[92] Negotiations between the two countries to lower the price were unsuccessful.[93] If Ukraine were to find a way to reduce its dependence on Russian gas, this would give Ukraine greater bargaining power when negotiating with Russia. Russia, on the other hand, would lose leverage over a critical transit state and likely see a reduction in its price-setting ability.

One way for Ukraine to reduce its dependence on Russian gas would be to increase its domestic production of gas. Two potential sources for additional Ukrainian gas production are the development of offshore gas fields in the Black Sea and onshore shale gas fields. In the aftermath of the 2009 gas crisis, Ukraine actively pursued both of these options. The northwestern Black Sea has the potential for significant gas production, and much of these reserves are off the coast of the Crimean Peninsula. In 2012, Ukraine accepted a bid from a consortium including ExxonMobil, Royal Dutch Shell, Romanian OMV Petrom, and Ukrainian Nadra Ukrainy to develop the Skifska gas field in the Black Sea.[94] The following year, Ukraine reached a separate deal with Italy's Eni and the France's EDF to jointly explore for oil and gas in a different area of the western Black Sea.[95] Once extracted, this offshore gas could be shipped from the Black Sea coast to domestic customers through Ukraine's existing pipeline network.

In addition, Ukraine has the fourth-largest shale gas reserves in Europe.[96] The largest of these reserves are the Yuzivska gas field in the eastern Donetsk and Kharkiv regions and the Olesska gas field in the western Lviv region. In 2012, Ukraine reached an agreement with Royal Dutch Shell to jointly develop the Yuzivska area. The following year it reached a similar deal with Chevron for production of the Olesska field.[97] With this potential increase in both shale and offshore gas production, Ukraine hoped to reduce its dependence on Russian

91. BP 2010.

92. Reuters 2013.

93. Reuters 2012.

94. Polityuk 2012.

95. Reuters 2013.

96. US Energy Information Association 2013b.

97. Reuters 2012.

gas and gain more leverage in negotiations with its neighbor. In contrast, Russia would benefit from maintaining its position as the primary supplier of gas to Ukraine. Additionally, Russia would want to prevent the emergence of a rival gas supplier in the region. Thus, Russia has strong incentives to take steps to prevent Ukraine from further developing its gas supply capabilities.

As was the case with Georgia, territorial expansion provided one avenue for Russia to attempt to achieve its market power goals. The unrest in Ukraine following the Euromaidan protests that began in late 2013 provided Russia with a window of opportunity to pursue a policy of territorial expansion in Ukraine. Russia attempted to increase its territorial footprint on two fronts. First, it moved to formally annex Crimea into the Russian Federation. Second, it intervened to support separatist forces in the eastern Ukrainian regions of Donetsk and Luhansk, collectively known as the Donbas. As with South Ossetia and Abkhazia, the separatist-controlled area within the Donbas has the potential to become the site of a prolonged frozen conflict, which would provide Russia with an opportunity to further expand its influence into Ukrainian territory.

Annexation of Crimea

Crimea is a strategically located peninsula on the north shore of the Black Sea. The Russian Empire had gained control of Crimea in 1783, and the peninsula remained part of Russia until 1954, when it was transferred to the Ukrainian Soviet Socialist Republic. Upon the breakup of the Soviet Union, Crimea remained part of the newly independent Ukraine. Russia, however, had designs on reincorporating Crimea into the Russian Federation. A majority of Crimean residents are ethnically Russian and speak Russian as their primary language. Critically for Russia, the port of Sevastopol in Crimea is the home of the Black Sea Fleet. In 1997, Ukraine agreed to a twenty-year lease in Sevastopol for the Russian Black Sea Fleet. Upon the accession of Yanukovych to the Ukrainian presidency in 2010, Russia and Ukraine signed the Kharkiv Accords, which extended Russia's lease in Sevastopol for an additional twenty-five years in exchange for a reduction in gas prices for Ukraine.[98]

The Euromaidan protests erupted in the response to the decision of Ukrainian President Yanukovych not to sign the European Union's Association Agreement. In December 2013, Yanukovych and Putin signed an action plan in which Russia would provide discounted gas prices and a $15 billion loan to Ukraine in return for a commitment by Ukraine to pursue closer ties to Russia. The protests escalated in

---

98. Smith 2016, 91.

early 2014, eventually leading Yanukovych to leave the country for exile in Russia on February 22, 2014. Within days, Russian troops entered Crimea.[99] Soon thereafter, Crimean authorities scheduled a referendum on the political status of the peninsula. The referendum resulted in the annexation of Crimea by Russia on March 18, 2014.[100] The annexation was widely condemned by most of the international community, and Western countries imposed sanctions on Russia for its actions. However, Russia proceeded to incorporate Crimea into the Russian Federation.

The annexation of Crimea enhanced Russia's strategic position in the Black Sea region. Russia now controls the critical port of Sevastopol, which houses the Black Sea Fleet. Additionally, absorbing Crimea potentially gives Russia access to additional maritime oil and gas reserves in the Black Sea. If Russia's incorporation of Crimea were ever to be recognized under international law, it would have the right to establish an exclusive economic zone within 200 nautical miles of the Crimean coastline. Even if Russia does not pursue the development of these offshore hydrocarbon resources, its actions in Crimea have blocked Ukraine from being able to do so. As we discussed earlier, in the years preceding the conflict Ukraine had signed deals with a consortium of energy companies to explore and develop offshore gas fields in the Black Sea. In response to the conflict in Ukraine, the companies decided to put these projects on hold.[101] Additionally, after the annexation, Russia seized control of the Crimean subsidiary of Naftogaz, the Ukrainian national gas company.[102] By annexing Crimea, Russia blocked Ukraine's ability to access oil and gas resources in the Black Sea and potentially opened the window for Russia to extract them itself in the future.

Donetsk and Luhansk

After the annexation of Crimea, the protests in Ukraine continued to intensify. Separatist groups in eastern Ukraine declared the independence of the Donetsk People's Republic (DPR) and the Luhansk People's Republic (LPR). In the spring and summer of 2014, armed conflict between these separatist groups and the Ukrainian government escalated, and Russia intervened to support the rebel groups with both arms and personnel. With this support, the DPR and the LPR were able to gain control of a significant part of the Donetsk and Luhansk

---

99. Smith 2016.

100. Myers and Barry 2014.

101. Reuters 2014a, 2014b.

102. Blockmans 2015.

regions. A pair of ceasefires (Minsk I and II) were reached by Ukraine, Russia, and the separatist groups in September 2014 and February 2015.[103] However, the conflict continued. Several additional ceasefire attempts over the following years have also failed. The conflict has largely reached a stalemate, but a low level of violence continues. If the violence dissipates, eastern Ukraine has the potential to become another frozen conflict on Russia's border.

As with South Ossetia and Abkhazia, Russia has largely pursued a policy of strategic delay in Donetsk and Luhansk. The status of these regions remains uncertain, as they are legally part of Ukraine but much of the contested area is under the de facto control of the pro-Russian DPR and LPR. While no settlement has been reached, Russia has also purposefully refrained from actively engaging in direct armed conflict with Ukraine. Instead, it indirectly challenges Ukraine using gray zone tactics like providing support to rebel forces. Moreover, Russia has generally preferred to focus on covert tactics and often officially denies any active involvement in the Ukrainian conflict. By perpetuating the property rights disputes in Donetsk and Luhansk, Russia is able to keep Ukraine in a strategically vulnerable position. Additionally, as long as these disputes remain alive, Russia leaves open the possibility of taking incremental steps to increase its influence in these regions, like it has in South Ossetia and Abkhazia.

Russia's strategy of strategic delay has also complemented its market power goals by interrupting Ukraine's attempts to produce more gas domestically. As noted earlier, eastern Ukraine is home to the large Yuzivska shale gas field. As a result of the conflict, Shell suspended its plans to develop this gas field in July 2014 and pulled out completely in the following year.[104] Due to the instability in the country, Chevron also ended its plans to develop the Olesska shale field in western Ukraine.[105] Given the halted projects in both the Black Sea and the shale gas fields, Ukraine was unable to increase its domestic gas production. In fact, as of 2018, Ukraine had not yet even been able to return to its 2014 levels of gas production.[106] However, Ukraine has taken steps to reduce its dependence on direct Russian gas imports by importing gas from EU countries to its west through reverse flow technology. Of course, by and large, this gas originally comes from Russia.

---

103. MacFarquhar 2014, 2015.

104. Olearchyk 2015.

105. Reuters 2014c.

106. BP 2019.

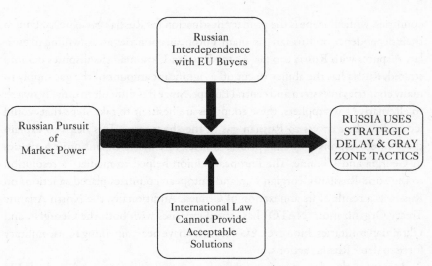

FIGURE 6.3. Russian market power and gray zone tactics.

Economic and Institutional Constraints

In the ongoing disputes over territorial control in Georgia and Ukraine, Russia has largely pursued a strategy of strategic delay. While it takes steps to prevent a permanent settlement of these property rights disputes, its actions also typically remain in the gray zone below full-scale war. By perpetuating these disputes and keeping the status of contested regions uncertain, Russia is able to exploit the vulnerability of its neighbors and keep its future options open. This allows Russia to engage in salami tactics that enable it to expand its reach without triggering a large-scale response from its rivals. While Russia's market power ambitions help to explain why it would want to pursue a policy of territorial expansion, to understand why it chooses to rely upon a strategy of delay, we also need to consider the constraints on Russia's behavior. According to our model of market power politics, strategic delay in the face of market power motivations is most likely to occur when an actor is constrained by economic interdependence but is willing to eschew an institutional settlement (see Figure 6.3). In the next section, we consider how economic interdependence constrains Russia and turn to the role of institutions in the section that follows.

Economic Interdependence

There is a high level of interdependence between Russia and the European Union in the regional gas market. Russia is a critical supplier of gas for many European countries, while Europe is the primary destination for Russia's gas exports. Due to their dependence on Russian gas, many European countries are unwilling to escalate disputes with Russia too far. As the Russian-Ukrainian gas disputes demonstrated, Russia has the ability to cut off a significant amount of the gas supply to many countries in eastern and central Europe. Since it is difficult to quickly switch to alternative gas suppliers, these countries are hesitant to take steps that would jeopardize their access to Russian gas in the short term. This helps to explain the limited responses by European countries to Russia's expansionist activities in Georgia and Ukraine. The European Union helped to mediate a resolution to the 2008 Russian-Georgian war, and European countries placed sanctions on Russia as a result of its annexation of Crimea. Additionally, the North Atlantic Treaty Organization (NATO) has staged exercises with both the Georgian and Ukrainian militaries. However, EU countries have been unwilling to use military force to deter Russia's actions.

However, this does not mean that there are no situations under which EU countries would respond militarily to Russian expansionism. If sufficiently critical interests are challenged, this could override concerns about Russian gas. For example, one would expect that EU (or NATO) members would intervene to defend against a direct attack on a fellow EU (or NATO) member. Thus, the EU response to Russia's activities likely depends upon the severity of the challenge to the interests of EU countries. The differing levels of European responses to the conflicts in Georgia and Ukraine can in part be explained by the variance in this source of economic constraint imposed on the EU buyers. At the time of the Russian invasion into Georgia, the European Union did not perceive Georgian territory as essential to the delivery of natural gas to EU markets. Ukraine, on the other hand, sits between the seller and the buyer in this scenario and remains an important cog of the supply chain. Thus, the EU reaction to the Russian incursion into Ukrainian territory was swifter and more intense. However, even in the case of Ukraine, EU countries relied upon the imposition of sanctions rather than the threat of military force and were not successful in reversing Russia's encroachments into Crimea, Donetsk, and Luhansk.

Of course, Russia is also highly dependent upon Europe. The vast majority of Russia's gas exports head to Europe, and the same infrastructural issues that prevent gas consumers from shifting to alternative suppliers in the short term also limit the suppliers from quickly accessing buyers in new markets. Additionally, Gazprom relies upon the proceeds from its European gas sales to subsidize domestic gas prices in Russia. Loss of these export markets would likely lead to increased energy prices within Russia, which could lead to domestic unrest. Thus, Russia cannot afford to take actions that would greatly risk a significant retaliation by EU countries, as this would increase European incentives to pursue alternative gas suppliers in the medium to long term. Rather, Russia would prefer to achieve its policy goals through a strategy that does not threaten European interests to such an extent that it sparks a significant escalatory response.

Thus, economic dependence decreases Russia's willingness to escalate its territorial disputes with Georgia and Ukraine militarily. Instead, it prefers to pursue a policy of strategic delay that allows it to engage in salami tactics to achieve its goals over time. With this strategy, Russia has slowed the pace of its expansion in order to avoid drastic and expensive efforts by EU buyers to exit their relationship with Gazprom. Each step taken by Russia is not sufficient to instigate a significant retaliatory response by European leaders. If Russia were to expand too quickly, EU states would likely find the domestic political will to pursue other energy options at any cost. On the other hand, if Russia fails to combat European Union attempts to increase the supply and diversify its suppliers of gas, it risks losing any market power it has accumulated thus far. The result is a need for Russian expansionism that is constrained by the current economic ties with the European Union. As such, Russia has strategically delayed resolutions to the property rights disputes that enable it to pursue a strategy of creeping territorial expansion.

Incompatible Institutions

By pursuing a policy of strategic delay, Russia has consciously avoided either a large-scale military escalation or an institutional solution to its property rights disputes with Georgia and Ukraine. The economic constraints provided by its dependence on Europe as an export market help to explain Russia's reluctance to significantly escalate these disputes. To understand why Russia also avoids reaching a lasting settlement to these disputes, we need to consider the nature of international institutions as they relate to these disputes. In particular, it is highly unlikely that Russia would be able to achieve a desirable outcome based upon the provisions of international law. At this time, Russia does not have a strong international legal claim for its occupation of South Ossetia and Abkhazia or its annexation of Crimea. The breakaway regions of South Ossetia and Abkhazia are only officially recognized as independent states by five countries (including Russia). Similarly, the vast majority of the international community has not recognized Russia's annexation of Crimea. Since it is likely that any international court would recognize South Ossetia and Abkhazia as Georgian territory and Crimea as Ukrainian territory, Russia has little incentive to pursue an institutional resolution of these territorial disputes.

Russia also benefits from delaying an internationally recognized settlement that defines the legal status of the contested territories. For one, there is the remote possibility that the international consensus on the issue may change, leading more countries to recognize the independence of the breakaway regions or Russia's annexation of Crimea. More importantly, delaying these settlements allows Russia to gradually shift the status quo in its favor by consolidating and expanding its de facto control over these territories. As we noted in Chapter 3, legal settlements of territorial disputes are often difficult to renegotiate. Thus, if Russia were to support a settlement finalizing the status of these disputed territories, it would take away its ability to continue expanding its territorial control through salami tactics.

In sum, our model of market power politics helps to explain Russia's use of gray zone strategies and salami tactics in Georgia and Ukraine. To protect its position in the European gas market, Russia has incentives to inhibit Georgia's ability to serve as a stable transit state for hydrocarbons from the Caspian region. Russia would also like to increase its bargaining power vis-à-vis its own primary transit state, Ukraine: Expanding its territorial reach provides one means by which Russia can help to achieve these goals. Given its dependence upon the European Union as its primary export market, Russia is constrained from escalating territorial disputes with its neighbors too far to avoid provoking a significant response by EU countries. Instead, it chooses to slowly expand its territorial reach through a series of smaller salami tactics that will not spark a significant response from European countries, many of whom are themselves reliant upon Russian gas imports. Additionally, given that the territorial claims of both Russia and pro-Russian separatist groups in Georgia and Ukraine are largely inconsistent with international law and are not recognized by most of the international community, Russia would prefer to avoid using international institutions to resolve these disputes. Instead, it generally chooses to pursue a policy of strategic delay.

Alternative Explanations

Two alternative explanations are worth exploring for Russia's use of gray zone tactics and piecemeal territorial expansion. The first common explanation for Russia's incursions into Ukraine has to do with its wariness of NATO expansion and the Ukrainian government's overtures toward NATO for potential membership. Given Putin's historical role during the Cold War era, it is logical to assume he viewed the NATO alliance warily and would actively resist Ukrainian membership. After all, if Ukraine were to obtain NATO membership, then the Alliance's Article V would trigger a collective defense in response to any Russian territorial incursions. While this view helps us understand why Putin would seek to thwart Ukraine's membership in NATO, it does little to explain the timing of Russia's invasion into eastern Ukraine and Crimea. The NATO threat has been relatively constant for over a decade. The recent discovery prior to the invasion of shale gas deposits in Ukraine, however, does provide a more focused motivation that fits the timeline of the invasion.

A second alternative explanation involves the domestic political arena in Russia. Like most states in the nascent twenty-first century, Russia is no stranger to nationalist movements. The annexation of Crimea and the occupation of Ukrainian territory appeals to such movements. Both Crimea and the regions of Donetsk and Luhansk are home to many Russian-speaking citizens. Yet it is difficult to envision a legitimate domestic challenger to the Putin regime as long as Russia is able to continue to garner wealth from its market power in hydrocarbons. Therefore, while it is clear that Russia's annexation of Crimea and support of separatist movements in eastern Ukraine are politically useful for Putin, it is hard to imagine that he needed these operations to succeed in order to remain in office.

Conclusion

Overall, Russia has had mixed success in its market power ambitions in the European gas market. Russia still remains the largest exporter of natural gas to Europe, and many countries do not have the ability to switch to alternative suppliers in the short term due to the existing pipeline infrastructure. Additionally, in the coming years, gas reserves in the European Union and Norway will continue to decline. However, many European countries are increasing their ability to import LNG from global suppliers, and pipelines transporting gas from Azerbaijan to Europe will soon go online. The emergence of these additional suppliers, along with an increase in hub pricing and intra-European gas trade, will potentially limit Russia's price-setting capabilities in the coming years.

Gazprom has been able to achieve some progress in vertically integrating the supply chain of gas from Russia to Europe. Through bilateral negotiations, the Russian gas company has been able to gain control of the gas transit infrastructure in Belarus, Moldova, and Armenia; however, similar attempts in Ukraine and Georgia were unsuccessful. The construction of the Nord Stream pipeline allows Russia to export gas directly to Germany without transiting through Ukraine. On the other hand, Gazprom had to scrap its plans to build the South Stream pipeline that would have bypassed Ukraine to the south. However, Russia's pursuit of the South Stream project likely contributed to the failure of the rival Nabucco project, which protected Russia from competition from Azerbaijani gas in parts of southeastern and central Europe. As a second-best alternative to South Stream, Gazprom has constructed the TurkStream pipeline. This has achieved Russia's goal of bypassing Ukraine, but it still requires Russian gas to pass through a transit state, Turkey, before reaching the European Union.

Similarly, Russia's use of strategic delay and salami tactics in territorial disputes in Georgia and Ukraine has yielded some gains but still leaves Russia short of maximizing its market power goals. Russia continues to consolidate its control over South Ossetia and Abkhazia, which remain outside the control of the Georgian government. Russia's continued military presence in these regions keeps Georgia in a perpetually vulnerable position. However, Georgia has largely refrained from importing Russian gas over the past decade, and it has maintained its position as a key transit state for oil and gas exports from Azerbaijan. Additionally, construction of the Southern Gas Corridor is still proceeding, and Azerbaijani gas will soon flow through Georgia on its way to customers in southern Europe.

Regarding Ukraine, Moscow has proceeded to incorporate Crimea into the Russian Federation, and pro-Russian separatists maintain de facto control over parts of Donetsk and Luhansk. Russia's presence in Crimea has prevented Ukraine from proceeding to access hydrocarbon resources in the Black Sea, and the conflict in Ukraine led Western companies to drop plans for shale gas production. However, while it has blocked Ukraine's plans, Russia has not been able to significantly capitalize on its access to additional gas reserves in the Black Sea, in part due to the ongoing economic sanctions imposed in the wake of its annexation of Crimea. Additionally, to reduce its vulnerability to Russia, Ukraine has moved to diversify its suppliers of natural gas and has largely stopped importing gas directly from Russia. Instead, it imports gas from EU countries using reverse flow technology. Of course, most of this gas is originally sourced from Russia. With Ukraine's diversification of its gas suppliers and Russia's moves to circumvent gas transit through Ukraine, both countries continue to take steps to reduce their dependence on each other in the gas trade.

These dynamics of the European gas market are likely to remain consistent, at least in the near term. Russia will continue to have incentives to maximize its market power by increasing its share of European gas imports, preventing the rise of alternative suppliers, and capturing control of the supply chain. At the same time, EU countries have incentives to increase competition between gas suppliers to prevent Russia from maintaining price-setting capabilities. Thus, we are likely to see further moves by Russia to further capture the market and countermoves by EU countries to forestall any Russian advances. The motivations of the players in the gas market will remain the same, even if their particular strategies shift in response to new challenges that emerge.

Given the current structure of both economic relations and international institutions, we do not expect a resolution in the near term of the "frozen conflicts" in South Ossetia and Abkhazia, the contested annexation of Crimea, or the stalemates in Donetsk and Luhansk. As long as European countries rely upon

Russian gas, they will likely remain reluctant to pursue significant retaliatory responses to Russian activities that do not critically threaten EU interests. Thus, EU countries may not be able to deter small-scale Russian incursions into their non-EU neighbors. At the same time, given its dependence upon Europe as an export market, Russia will likely refrain from significant expansionist behavior that would provoke European actors to make accelerated moves to exit their relationships with Russia. Thus, Russia will likely continue to pursue salami tactics that advance its interests but do not threaten military escalation. At the same time, under current international rules we are unlikely to see Russia willing to seek institutional settlements to the territorial disputes in the region. Instead, Russia will likely continue its practice of strategic delay.

However, there is always the possibility that at some point Russia may be willing to reach a settlement in these property rights disputes. The successful resolution of the maritime dispute in the Caspian Sea illustrates how this may come about. In this case, disputed boundaries had a significant bearing on Russia's position in the natural gas market, and Russia eventually transitioned from a policy of strategic delay to a negotiated settlement. After over two decades of strategic delay by Russia, the five bordering states finally reached an agreement on the legal status of the hydrocarbon-rich Caspian Sea in 2018. We will consider this case and its implications for successful dispute settlement in the face of market power opportunities in concluding chapter of the book.

Before we discuss the successful navigation of the disputes surrounding the Caspian Sea, however, in the next chapter we turn to a set of maritime disputes that involve China and its neighbors to the east and the south. Just as Russia has embraced gray zone tactics to pursue expansionism and disruption in pursuit of market power goals, China has energetically pushed its claims of sovereignty in the East and South China Seas. We will argue that an underlying factor driving China's expansionism is the pursuit and maintenance of market power, particularly in the global rare earth elements market. Why has this expansionism been more dramatic and aggressive in the South China Sea compared to the East? Our analysis suggests that China has been held in check by Japan in the East China Sea, whereas none of the neighboring states in the South has been able to leverage political or economic tools to slow China down.

7

China

CAPTURING SEABED RESOURCES

IN EARLY SEPTEMBER 2010, a Chinese fishing trawler collided with two Japanese patrol boats in disputed waters near the Diaoyu/Senkaku Islands in the East China Sea. The Japanese Coast Guard arrested the captain of the trawler after he repeatedly ignored warnings to leave and subsequently collided with each patrol boat. His arrest sparked outrage from the Chinese government and its civil society, and China's Foreign Ministry called for his release and a cessation against any "so-called law enforcement activities or any actions that would jeopardise Chinese fishing boats or Chinese people."[1] The event catalyzed what would become known as the Rare Earth Crisis of 2010–2012, in which the world witnessed skyrocketing prices for rare earth oxides.[2] Prices spiked as much as 2,000 percent for some elements.[3] By 2012, however, both the political and economic crises seemed to be mitigated, if unresolved formally, and prices stabilized. In terms of the absence of escalation to a violent military dispute, the 2010 incident is not unique. At least a half-dozen such incidents have occurred between the People's Republic of China (PRC) and Japan over these islands, and each time the two states manage to step back and avoid conflict while delaying the resolution of the dispute.

China's maritime disputes in the South China Sea, however, tell a different story. In early 2013, the Philippines began arbitration proceedings against China with respect to maritime jurisdiction in the South China Sea over the Scarborough Shoal and Spratly Islands. In response, China issued a third-party

---

1. BBC News 2010, n.p.

2. Cox and Kynicky 2018, 1.

3. Klinger 2018.

note repeating its maritime claims and declined to participate in the arbitration. Instead, China has continued to build military bases on artificial islands in this area. According to a 2017 US Department of Defense report, by the end of 2016 China had *added* over 3,200 acres of land and had placed airfields, hangars, military barracks, fixed weapons positions, and fuel depots on these makeshift bases. China is rapidly expanding its military presence in the region and is establishing de facto sovereignty over islands more than 700 miles away from the mainland.[4]

What explains the duality of these strategies? In the East China Sea, China has repeatedly backed away from aggression, instead relying on treaties and negotiations with Japan to dampen conflict. In the South China Sea, China has been far bolder, building up its military presence with little regard to international treaties or the outcries of its neighbors. In both cases, the legal maritime disputes over the sovereignty of these areas remains unresolved, and China shows no desire to engage with the international institutions that can provide a resolution. Instead, China seems content to delay any resolution at all as it slowly implements a "cabbage strategy"[5] of salami tactics to bolster its de facto position.

Both sets of events, along with a myriad of similar events in the South and East China Seas, can be best explained by the strengths and limitations of China's market power and market power goals. Here we will focus on the specific market of rare earth elements (REEs) and the role of these elements in China's broader Made in China 2025 economic strategy.[6] While the origins of China's interest in holding market power over REEs dates to the early 1990s, this goal became a key component of Chinese economic strategy in 2015 with the rollout of the Made in China 2025 blueprint. China's economic goal is to be the first choice for producers using these minerals throughout the global value chain while maintaining stable and competitive access to rare earths for China's domestic economy.

Ideally, international institutions would provide an avenue for China to solve these issues. In both the East and South China Seas, the legal institutions provided by the United Nations Convention on the Law of the Sea (UNCLOS) and the International Seabed Authority (ISA) provide a framework for dispute resolution.[7] However, while China has been an active member of the UNCLOS regime and its policies generally, it has refused to engage in the legal process to

---

4. United States Department of Defense 2017.

5. This strategy involves wrapping layers of sovereignty around an island like leaves of cabbage.

6. We are deeply grateful to Krista Wiegand for pointing us in this direction.

7. The International Seabed Authority is an autonomous organization established under the authority of the United Nations Convention on the Law of the Sea. States that are parties to the Convention are parties to the ISA, which specializes in the seabed and what lies underneath. The rest of the UNCLOS regime applies to the water, its surface, and the airspace above.

resolve these disputes. This reluctance is due to the incompatibility between the design of these institutions and China's market power goals in the East and South China Seas. As we will see in the following discussion, the international institutions work well for what they were designed to accomplish. They simply do not provide a dispute-resolution option that allows China to maintain unfettered access to deep sea mineral extraction, a key component of China's rare earth strategy.

While the design of international institutions has made China reluctant to resolve these disputes in a legal forum, China's economic relationships place constraints on willingness to use military force to achieve its market power goals. In the East China Sea, China's dependence on Japanese trade and investment prevents China from escalating this dispute too far. On the other hand, China's asymmetrically interdependent relationship with its South China Sea neighbors has allowed China to take more aggressive steps to the south. However, China's overall dependence upon the global REE market constrains China from using excessive military force to seize control of resources in the South China Sea.

As a result, China has turned to the strategies of strategic delay and gray zone tactics to continue to push its claims for sovereignty in the East and South China Seas. The need to preserve its market power prevents China from embracing the resolution options available through UNCLOS. At the same time, the need to maintain a stable market price and deter new entrants prevents China from more volatile military action. Like Russia, China is likely to continue this strategy, at least until it no longer benefits from holding this market power. While the need for rare earths is not the sole motivation for controlling these waters, similar dynamics surround the oil and gas reserves in the region. China appears to be committed to a strategy of strategic delay, as it leverages its access to REEs to become competitive in renewable energies and advanced electronics manufacturing.

Our goal in this chapter is to investigate China's pursuit of market power in the East and South China Seas to help us understand its continued use of strategic delay. In the pages that follow, we will explore the rich history of interactions among the neighbors in the East and South China Seas and introduce the reader to the hard commodities that are REEs. We will then examine China's motives for creating and maintaining market power, as well as Japan's sustained efforts to prevent it. Then, to understand China's use of strategic delay in pursuit of these goals, we investigate how the design of international institutions like UNCLOS and the ISA has made China less willing to settle its maritime disputes and how the structure of economic interdependence has constrained disputants from turning to military escalation. Lastly, we evaluate alternative explanations, particularly the domestic and international reputation constraints that prevent compromise, as well as the notion of controlling the open waters in a bid to establish control in the region and restrict access.

China's Disputes in the East and South China Seas

For decades, China has faced ongoing property rights disputes in the East and South China Seas. In both seas, the disputes involve a combination of competing territorial and maritime claims. The territorial disputes center around the sovereign control of small islands, reefs, and rocks, most of which have historically been uninhabited. For the most part, these islands are not significantly valuable in and of themselves. Instead, states value the possibility of controlling the seas and seabeds that surround the islands. Thus, these territorial disputes are inherently linked to related maritime disputes over the location of states' territorial seas and exclusive economic zones (EEZs), which affects the ability of states to access marine resources in these areas.[8] These territorial and maritime disputes have received a higher profile in recent years as China has continued to overtly press its territorial and maritime claims in the East and South China Seas. Let us take a closer look at China's activities in these areas.

Restraint in the East China Sea

The long-standing dispute between China and Japan in the East China Sea involves competing territorial claims over the Diaoyu/Senkaku Islands and a disagreement over their maritime boundaries. Both countries claim sovereignty over the island chain, which the Chinese call Diaoyu and the Japanese call Senkaku.[9] Located about 200 nautical miles from both mainland China and the Japanese island of Okinawa, the disputed archipelago consists of five uninhabited islets and three rocks. In addition to this territorial dispute, the two countries disagree about the location of their respective EEZs. Based upon its interpretation of the extension of its continental shelf, China claims that almost all of the East China Sea falls within its EEZ. Japan, on the other hand, asserts that the states' mutual maritime boundary should be drawn based upon the median line between the

---

We will discuss the ISA and UNCLOS as two different regimes here, despite these connections, simply for clarity of exposition.

8. A territorial sea is considered part of a state's sovereign territory, while an EEZ is a maritime area in which a state has jurisdiction over the exploration and exploitation of marine resources.

9. While our analysis focuses on the dispute between China (PRC) and Japan, the islands are also claimed by Taiwan (Republic of China).

countries' coastlines. As the map in Figure 7.1 shows, these overlapping claims leave a significant region in which China and Japan disagree over who should have jurisdiction over the exploration and exploitation of maritime resources.

The Sino-Japanese dispute over the Diaoyu/Senkaku Islands has roots dating back to the fourteenth century, and over this time both China and Japan have

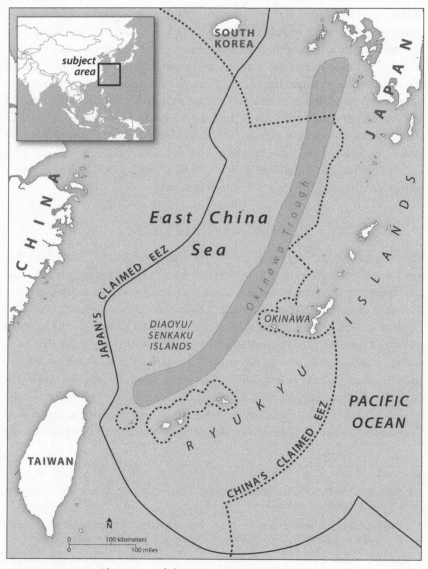

FIGURE 7.1. East China Sea and the Diaoyu/Senkaku Islands.

voiced claims to the islands and their surrounding waters.[10] At first glance, these islands may seem to offer no concrete gain to either country. The formations are rocky, uninhabitable, and merely the result of volcanic activity 200 nautical miles out into the East China Sea. However, the discovery of potential oil and gas reserves in the area in the late 1960s revealed the implicit value of the islands and sparked the modern-day territorial dispute between China and Japan.[11] In the period after World War II, the United States administered the Diaoyu/Senkaku Islands as part of the Ryukyu Islands. Then, in 1971, the United States and Japan signed the Okinawa Reversion Treaty, which transferred control of the Diaoyu/Senkaku Islands back to Japan the following year. In response, China made its first formal modern claim to the Diaoyu/Senkaku Islands on December 30, 1971, asserting that Japan was encroaching on its sovereign territory.[12] Amidst Chinese claims, the transfer of power from the United States to Japan marked the beginning of a cyclical pattern in increasing and decreasing tensions between the Chinese and Japanese governments.

While we do not provide a detailed history of the Diaoyu/Senkaku dispute here, the flare-up in 2004–2005 and the 2010 trawler incident illustrate the nature of the strategic interactions between China and Japan over this issue. In the early 2000s, competition between China and Japan over the development of hydrocarbons in East China Sea increased as both countries took steps to explore for natural gas in the area around the median line claimed by Japan as its EEZ boundary.[13] This resource competition coincided with an increase in intrusive tactics by China, including the landing of Chinese activists on one of the disputed islands in March 2004 and the deployment of Chinese naval ships and a submarine to the area in the fall of that year. Then, in April 2005, violent anti-Japanese protests involving tens of thousands of citizens broke out in major cities across China.

As the two countries sat down for talks on the Diaoyu/Senkaku dispute in September, China deployed additional naval ships and spy planes to the area.[14] At these talks, Japan proposed joint development of gas fields around the median line, which China initially rejected because it feared such a concession would signal a tacit approval of Japan's EEZ boundary claim.[15] China then canceled

---

10. We are grateful to Rebecca Kalmbach for contributing to the research on this history.

11. Shaw 1999.

12. Wiegand 2009.

13. Tsai 2016.

14. Wiegand 2009.

15. Valencia 2007.

diplomatic talks with Japan in October to protest Prime Minister Koizumi's visit to the Yasukuni Shrine. Talks resumed in the following year, and in June 2008, the two countries signed a joint natural gas development agreement. However, neither side made any concessions with respect to their territorial and maritime claims.[16]

Just over two years later, one of the most explosive incidents occurred between agents of both states in the 2010 trawler incident, often referred to as the 2010 Senkaku Crisis. On September 7 of that year, a Chinese fishing trawler encountered three patrol vessels from the Japanese Coast Guard (JCG). The JCG reported that they discovered the Chinese fishing vessel approximately twelve kilometers north of the islands and ordered the vessel to stop for inspection. The captain, Zhan Quxiong, of the Chinese vessel refused the order and in the ensuing pursuit the fishing trawler collided with two of the three coast guard boats. Once overpowered, Captain Zhan and his crew were detained by the JCG. Perhaps the most inflammatory moment came when the Japanese government declared that Zhan should be tried under domestic Japanese law for obstructing the Coast Guard.[17]

The response from China was immediate, starting with the cessation of East China Sea negotiations on September 10. Although Japan released the crew quickly, it extended the detention of Captain Zhan and continued its investigation. China's reaction again escalated, detaining four Japanese citizens in China. But the Chinese response that caused the greatest amount of concern was its decision to embargo the export of rare earths to Japan. At that time, Japan imported over 80 percent of its rare earths from China, and these metals were (and are) extremely important to Japanese electronics manufacturing. China also announced a quota system for limiting rare earth exports to the entire market.

On September 29, Japan abruptly released the captain, citing the diplomatic impact of the crisis, but it would be three months before China lifted its embargo. Both Japan and the United States responded to China's economic statecraft with panic, holding high-level security briefings, securing national security testimony in the US Senate, and launching new investigations into shifting Japan's rare earth dependence away from China. Tense exchanges between China and Japan would continue well into October, and Japan's domestic political arena wrestled with the event through the spring of 2011.

The last fifty years are replete with examples of this pattern. Both China and Japan have allowed heated domestic discontent toward one another's claims,

16. Wiegand 2009.

17. BBC News 2010; Smith 2012.

dismissed each other's positions, and in some instances used military might to create tension in the area. Each time, both sides have backed away from the possibility of significant militarized violence and followed each flare-up with public demonstrations of cooperation and tolerance. However, once these tensions die down, China and Japan have never taken the next step of trying to settle the underlying territorial dispute. Instead, the two states continue to maintain their competing claims of sovereignty over the islands. In short, both states have opted to strategically delay any resolution of the Diaoyu/Senkaku dispute.

Aggression Unchecked in the South China Sea

As in the East China Sea, China's ongoing property rights disputes in the South China Sea consist of competing maritime and territorial claims. China claims jurisdiction over the area that falls within the so-called nine-dash line, which constitutes about 90 percent of the South China Sea (see Figure 7.2). This area overlaps with the EEZs claimed by all of the other states surrounding the sea. In addition to disagreements over maritime boundaries, China's assertion of the nine-dash line has also led to territorial disputes over islands and reefs in the sea, including the Spratly Islands, the Paracel Islands, and the Scarborough Shoal. Six countries surrounding the sea—Brunei, China, Malaysia, the Philippines, Taiwan, Vietnam—claim sovereignty over all or part of the Spratly Islands. China, Taiwan, and Vietnam dispute control over the Paracel Islands, while China, the Philippines, and Taiwan make claim to the Scarborough Shoal.

While China bases its claim in the South China Sea on events that date back centuries, the dispute has its origins in the mid-twentieth century.[18] In 1950, China occupied Woody Island, the largest of the Paracel Islands. The following year, the PRC made its first official claim to sovereignty over the Paracel and Spratly Islands and expanded the claim in 1958 to include the surrounding territorial waters.[19] In the 1970s, exploration for potential offshore hydrocarbon reserves led to increased interest in the South China Sea. To bolster their claims to any of these resources, the surrounding states expanded their efforts to occupy islands and reefs in the sea. This competition over territory has twice escalated to military conflict. In 1974, a military clash between Chinese and South Vietnamese forces resulted in Chinese occupation of the entire Paracel archipelago. Over a decade later, in the wake of China's decision to occupy Fiery Cross Reef in the Spratly

---

18. We are grateful to Michael Purello for contributing to the research on this history.

19. Dupuy and Dupuy 2013; Fravel 2008.

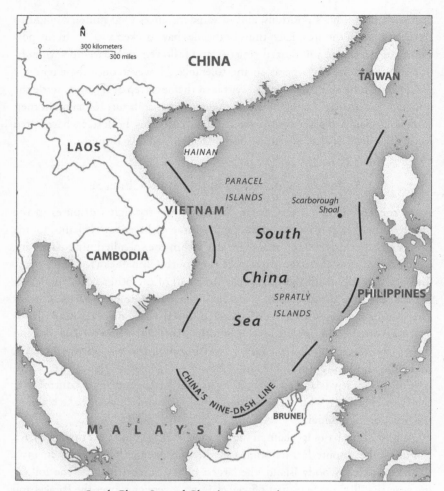

FIGURE 7.2. South China Sea and China's maritime claim.

Islands, another military clash broke out between China and Vietnam in 1988, resulting in the sinking of three Vietnamese ships and seventy-four casualties.[20]

Other than these two violent clashes with Vietnam, China has largely pursued a policy of strategic delay in the South China Sea. To explore this delay strategy, we will mainly focus our attention on China's high-profile dispute with the Philippines over the Spratly Islands and the Scarborough Shoal. However, the strategic dynamics we explore are similar to those found in

---

20. Fravel 2008.

China's disputes with the other claimants.[21] The Philippines occupied eight features in the Spratly Islands in the 1970s and formally declared sovereignty over the island group in 1978.[22] To counter the increased presence of the Philippines and Vietnam and bolster its own claims, China moved to occupy six reefs in 1988. Then, in 1995, China occupied Mischief Reef, one of the easternmost features in the Spratly Islands near the Philippines, and built four platforms on stilts that housed several bunkers equipped with a satellite dish to communicate with the mainland.[23] In 1998, China expanded and fortified these structures with the support of armed military supply ships.[24] Despite protests from the Philippine government, China has continued to maintain effective control of Mischief Reef. On-and-off-again negotiations between the China and the Philippines in the 1990s and early 2000s failed to settle their territorial and maritime disputes.

By the 2010s, China had begun to more aggressively press its claims in the South China Sea. The 2012 Scarborough Shoal incident illustrates China's more assertive stance against the Philippines. Located 124 nautical miles west of the Philippines, the Scarborough Shoal contains valuable fishing grounds. On April 10, 2012, a Philippine naval frigate intercepted eight Chinese fishing vessels it claimed was engaging in illegal activity. Two Chinese surveillance ships arrived and blocked the Philippine ship from accessing the fishing boats in the shoal's lagoon.[25] A two-month standoff between the two states ensued. In May, China dispatched four additional surveillance ships, and at one point there were ninety-seven Chinese vessels in the area around the shoal.[26] In addition to these gray zone tactics, China also imposed economic and diplomatic sanctions on the Philippines.[27] The specific details of the negotiations leading to the end of the standoff are disputed. However, in the end, the Philippines withdrew all of its ships on June 15, while the Chinese ships remained. China gained de facto

---

21. The main exception to this would be Taiwan, which has a more complicated political and economic relationship with the PRC given that the PRC claims that Taiwan is part of China. Given this added complexity, we set aside any discussion of the competing claims between China and Taiwan for this analysis.

22. Bautista and Schofield 2012.

23. Storey 1999.

24. Zha and Valencia 2001.

25. Whaley 2012.

26. Green et al. 2017.

27. Zhang 2019.

control over the Scarborough Shoal and has regularly stationed coast guard ships to restrict the activities of Filipino fishermen in the area.[28]

Along with the aggressive activities of the Chinese Coast Guard, another key part of China's expansionist strategy in the Spratly Islands involves the construction and fortification of artificial islands. In late 2013, China began land reclamation at Johnson Reef. Over the next two years, it engaged in dredging activities at all seven of the Chinese-controlled reefs in the Spratly archipelago. Once these land reclamation efforts were completed in mid-2015, China moved forward with construction efforts on the islands.[29] On all seven features, China built administrative structures and communications towers and installed defensive weapons systems. China has focused its most significant development activities on three of these islands—Fiery Cross Reef, Subi Reef, and Mischief Reef—which all have runways and hangars for combat aircraft.[30]

In January 2013, the Philippines initiated arbitration proceedings against China in the Permanent Court of Arbitration (PCA) to resolve the dispute over the two states' maritime claims under the authority of provided by UNCLOS. China asserted that the PCA lacked jurisdiction, and it refused to participate in the arbitration proceedings. The case, however, continued despite China's absence. The PCA issued its ruling in July 2016, finding that China's claim to historic rights to resources within the nine-dash line has no basis in international law under UNCLOS. Moreover, the PCA ruled that the disputed Spratly Islands and the Scarborough Shoal were either "rocks" or low-tide features and thus do not entitle a state to an EEZ or continental shelf claim. Finally, the court found that China's activities near Mischief Reef and Scarborough Shoal were illegal, given that both features lie within the Philippines' EEZ. Not surprisingly, the Chinese government immediately stated that it neither accepted nor recognized the ruling.[31]

Rather than taking a step back in the wake of the unfavorable arbitration decision, China has maintained an aggressive stance to protect its claims in the South China Sea. It has continued it efforts to militarize the Spratly Islands. By mid-2018, the Chinese military had quietly installed communications jamming equipment, anti-aircraft guns, anti-ship cruise missiles, and surface-to-air missile systems on Fiery Cross Reef, Subi Reef, and Mischief Reef.[32] The port facilities at these artificial islands also allow the Chinese Coast Guard to continue

---

28. Green et al. 2017.

29. Ibid.

30. Asia Maritime Transparency Initiative 2018.

31. Reed and Wong 2016.

32. Macias 2018.

to aggressively patrol the South China Sea and extend China's area of control beyond these currently occupied islands.[33] Chinese ships also regularly engage in efforts to harass Vietnamese and Malaysian oil and gas activities in the South China Sea.[34] While largely nonviolent, such patrols have the potential to escalate, as when a Chinese vessel rammed and sank a Vietnamese fishing boat near the Paracel Islands in April 2020.[35]

Putting this all together, we can see that China has used strategic delay as an opportunity to engage in gray zone activities to expand and consolidate its presence in the South China Sea. Through a series of salami tactics, China has moved slowly but determinedly in order to avoid a panic that could accidentally trigger an escalation in violence. Some observers have coined this policy a "cabbage strategy," as it consists of China "consolidating control over disputed islands by wrapping those islands, like the leaves of a cabbage, in successive layers of occupation and protection formed by fishing boats, Chinese Coast Guard ships, and then finally Chinese naval ships."[36] Rather than turning to international institutions to resolve these disputes over territorial and maritime jurisdiction, China appears to want to gradually shift the status quo in its favor until it has de facto control over the South China Sea.

Two Paths of Strategic Delay

The ongoing territorial and maritime disputes in the East and South China Seas thus illustrate the use of strategic delay, but with important differences. In the East China Sea, China and Japan have both adopted a strategic delay approach at different points over the last five decades. Early in the PRC's regime, both states eschewed violence in favor of negotiations that avoided the interruption of trade. In the last thirty years, however, China has expanded its maritime claims while Japan seeks to use strategic delay tactics to prevent China from deepening its monopolistic power over hard commodities in the region. Even as it makes these expansionist claims, though, China has largely refrained from taking aggressive actions against Japan to assert control over the East China Sea.

No such counterbalance exists in the South China Sea, allowing China to complement its policy of strategic delay with gray zone tactics as it moves forward

---

33. Asia Maritime Transparency Initiative 2019a.

34. Asia Maritime Transparency Initiative 2019b; Latiff and Ananthalakshmi 2020.

35. Vu 2020.

36. O'Rourke 2018, 33.

with its expansionist goals to control the resources of the South China Sea. China has continued to expand its grasp in the region by building islands and expanding its military presence. Yet despite this steadfast push, China has yet to allow any of the incidents surrounding this expansion to escalate to militarized violence. Rather than pushing hard for complete dominance in the South China Sea, China seems content to slowly wrap leaves of sovereignty around reefs, islands, and shoals as it inexorably pushes for the recognition of its historical claim. The result is a set of gray zone tactics driven by China, none of which has escalated to traditional conflict.

Previous studies have examined reasons for China to pursue strategic delay in the East and South China Seas. For example, Wiegand effectively argues that China has used the dispute over the Diaoyu/Senkaku Islands as leverage in other negotiations with Japan, extracting concessions in economic and political dimensions of their relationship.[37] China benefits from keeping the dispute alive, as it allows China to continue to use such issue linkage to compel Japan when disagreements arise. In his analysis of the South China Sea dispute, Fravel argues that China has pursued a delay strategy to consolidate its own claims in the sea and deter other states from strengthening their claims.[38] Given that offshore islands are relatively cheap to dispute and can be important for establishing the extent of a country's EEZ under the UNCLOS, China has incentives to bargain hard and wait for an optimal settlement.[39]

To understand China's use of strategic delay in these disputes and the diverging paths it has taken in the East and South China Seas, we turn to our theory of market power politics. While analysis here is not necessarily incompatible with these previous studies, it puts forward a novel configuration of motivation and constraints that shapes China's activities in its neighboring seas. In particular, we identify China's power in the REE market as a fascinating and troublesome motivation for China's foreign policy behavior. The presence of valuable seabed resources, including REEs and hydrocarbons, motivates China to press its claims in the East and South China Seas. Here we will mainly focus on the role of rare earths, but the motivation is consistent for oil and gas as well. As the largest global producer of REEs, China has the ability to set prices in the market. Access to deep-sea rare earth deposits would help China preserve its market power and prevent other states from gaining the power to affect the prices of REEs.

---

37. Wiegand 2009.

38. Fravel 2011.

39. Fravel 2008.

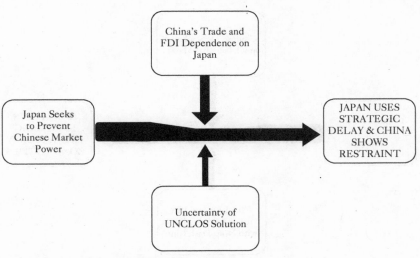

FIGURE 7.3. Strategic delay in the East China Sea.

In the East China Sea, Japan would also like to have access to these resources for its own use. But, more importantly, Japan wants to avoid any outcome that would provide China with a discontinuous spike in market power with respect to REEs. To this end, Japan has utilized its key role as a contributor of foreign direct investment (FDI) to China as an interdependence constraint that prevents China from escalating to more aggressive tactics. At the same time, the limitations of international institutions fail to provide a suitable path toward dispute resolution. Due to this configuration of motivations and constraints, China and Japan have pursued a practice of strategic delay in which they avoid reaching a permanent dispute settlement but also refrain from significant military escalation. Figure 7.3 shows how the three dimensions of market power opportunity, interdependence, and institutions combine to explain the use of strategic delay in the East China Sea.

Economic constraints work differently in the South China Sea dispute. Given the asymmetrical interdependence between China and the Philippines, China is not significantly constrained by its economic relationship with its South China Sea neighbor. However, concerns about adaptation by other global producers prevents China from turning to military escalation to press its claims. At the same time, international institutions like UNCLOS and the ISA fail to provide a dispute resolution option that allows China to maintain unrestricted access to these deep-sea resources. Figure 7.4 illustrates how this combination of motivation and constraints has led China to pursue a policy of strategic delay in the South China Sea dispute. However, in contrast to the restraint China has exercised in the East

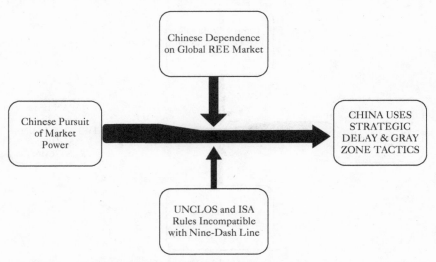

FIGURE 7.4. Strategic delay in the South China Sea.

China Sea, the asymmetrical economic relationship with the Philippines allows China to engage in more aggressive gray zone tactics in the South China Sea.

Later in this chapter, we will examine in more detail how these motivations and constraints shape the strategies of China and its fellow disputants in the East and South China Seas. Before doing so, however, we want to take a step back and provide an introduction to the global market for REEs. As we will see, China's dominant position in this market is a key component of its current and future economic strategy.

China's Market Power in the Rare Earths

Discussions of China's maritime claims often focus on the historical roots or China's geopolitical ambition to establish a sphere of influence in the Pacific Rim. While these claims are certainly complex and involve multiple dimensions of incentives, a sometimes overlooked motivation for China's ongoing interest in the East and South China Seas is the economic value of controlling the extraction of seabed resources. One such key economic dimension is the extraction of hard commodities from the seabed, such as oil, gas, and minerals. Here we focus on one particular set of commodities, REEs.[40] China has operated as the dominant player in the rare earth market since the 1990s, and it has had a near monopoly

---

40. We are grateful to Maya Schroder for her assistance with this research.

on the export of rare earths, which are widely being used in the manufacturing of advanced electronics. In this section, we provide an overview of these metals and discuss how China emerged as the dominant player in the REE market. We then delineate China's motivation for continued control of this market and set the stage for demonstrating why this motivation plays a key role in the maritime disputes that continue to roil the East and South China Seas in the twenty-first century.

What Are Rare Earth Elements?

Before we make the case that China's pursuit of stability in its power in the REE market is key to its geopolitical strategy in the East and South China Seas, let us take a moment to introduce these hard commodities. While the name suggests an elusive product, in truth it is likely that you have rare earths in the phone in your pocket, the watch on your wrist, and in nearly every battery-run or advanced electronic device you own. Rare earths are often considered the vitamins of modern industry, enabling advances in manufacturing devices as well as the manufacturing process. The elements themselves comprise seventeen metals, often divided into light and heavy classifications based on their atomic number.[41] Fifteen of the metals are classified as lanthanoids, due to shared physical and chemical properties. The last two, yttrium (Y) and scandium (Sc), while not lanthanoids, have some properties in common with the group. Together, the seventeen metals are collectively referred to as REEs.[42]

The production of REEs into refined metals is a costly and complex process, which we simplify here in order to illustrate the infrastructure needed. Mining is the first step, whereby the raw elements are extracted from the earth. While REEs are not very rare, they do tend to be sparsely distributed, requiring large amounts of mining in order to obtain small amounts of REEs. Once mined, the REEs must be separated from other metals, mud, and other substances, and then enter a chemical process to convert the raw ores into oxides and finally into refined metals. Currently, China is the only state with the large-scale capacity to convert REEs into refined metals suitable for manufacturing.

The REE market is growing, with growth in the global market estimated at nearly 14 percent yearly between 2017 and 2021. The size of the actual REE market

---

41. These seventeen metals include: lanthanum (La), cerium (Ce), praseodymium (Pr), neodymium (Nd), promethium (Pr), samarium (Sm), europium (Eu), gadolinium (Gd), terbium (Tb), dysprosium (Dy), holmium (Ho), erbium (Er), thulium (Tm), ytterbium (Yb), lutetium (Lu), yttrium (Y), and scandium (Sc).

42. Ganguli and Cook 2018; Seaman 2010, 2019.

is relatively modest, with an approximate value of $9 billion. This value is deceptive, however, because these refined metals are considered essential to manufactured goods worth an estimated $7 trillion yearly.[43] The downstream economic effects of the REE market are immense.[44] Whoever controls the production and exports of REEs will have a major impact on the global economy, affecting clean energy industries such as solar power and battery-powered vehicles, advanced electronics, permanent magnets, UV-repelling glass, air-conditioning equipment, and even chemotherapy medications.[45] As the world inches toward non-fossil-fuel energy sources, the importance of REEs in the manufacturing of batteries will become even more important. Military applications are also relevant, as REEs play an important role in the manufacturing of jet engines, satellites, missile guidance systems, communication devices, and night goggles.[46] For these reasons, there are significant benefits for a state that has the ability to exercise control over the REE market.

The Emergence of Chinese Market Power

Until the mid-1980s, the United States was the dominant producer of rare earths, and it continues to hold some of the largest known deposits of REEs. China entered the market in the late 1980s with lower costs in the manufacturing of refined metals. At the same time, US firms began to exit the market as they faced increasing costs due to domestic environmental regulations. China quickly established itself as the dominant global producer of REEs. In 1992, Deng Xiaoping is quoted as stating, "The Middle East has its oil. China has rare earths. . . . It is of extremely important strategic significance; we must be sure to handle the rare earth issue properly and make fullest use of our country's advantage in rare earth resources."[47] By the early 2000s, China extracted and refined nearly 97 percent of the rare earths purchased globally. Figure 7.5 illustrates this evolution of China's market position as a producer of REEs.

With a near monopoly on production, China gained the ability to set prices in the rare earth market and sought ways to use this market power to its benefit.

---

43. Ganguli and Cook 2018, 1.

44. Machacek and Fold 2018.

45. Ganguli and Cook 2018; Machacek and Fold 2018; Seaman 2019; Vekasi 2014.

46. Vekasi 2014, 150.

47. Ganguli and Cook 2018, 10. This quote is widely used, but difficult to substantiate with an original source.

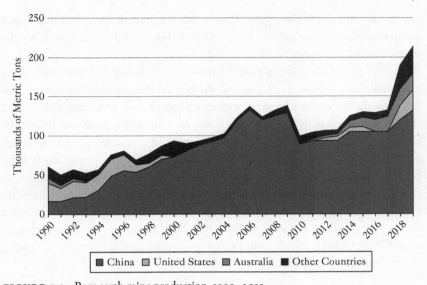

FIGURE 7.5. Rare earth mine production, 1990–2019.

Source: United States Geological Survey (USGS), *Mineral Commodities Summaries* and *Minerals Yearbook*, 1990–2020.

The 2010 fishing trawler crisis with Japan is often identified and discussed as a turning point in assertiveness in Chinese foreign policy, as well as its willingness to use its dominance of the REE market as a tool of economic statecraft. After the trawler incident, China rolled out quotas and taxes on its exports of REEs and their processed versions (rare earth oxides) and imposed a two-month ban on any exports of the minerals to Japan. Prices of REEs spiked, causing a major panic in the global markets. As Klinger wrote in 2018, "In late 2010, the world awoke to its dependence on China for 97 percent of the global supply of rare earth elements."[48]

Interestingly, however, it also appears that China had already planned to institute quotas and restrictions, and their imposition subsequent to the trawler incident was more a matter of timing than a change in strategy. China's Ministry of Industry and Information Technology released a draft of the plan in 2009, sparking diplomatic worry and financial interest.[49] The embargo against Japan, along with the broader management of the arrest of the Chinese fishing captain, seemed to be more motivated by domestic political pressures than a desire to

---

48. Klinger 2018, 2.

49. Cox and Kynicky 2018.

assert new influence in the region.[50] Given this domestic pressure, it seems almost surprising that tensions did not rise to violence. Instead, China and Japan rapidly de-escalated.

By the end of the crisis in 2012, Chinese market power had *weakened*, albeit marginally, as countries like Japan and the United States poured billions of dollars into their nascent rare earth extraction capabilities and Australia entered the market. Even as its market share dropped, China still produced four times the REEs than all other producers combined. China learned from this experience that spikes in the market price of REEs would stimulate market entry from powerful economies such as Japan and the United States, as well as new entrants from young mining firms with global investors.[51] The rapid adaptation by other players caused China to retreat from its position with Japan as soon as domestic politics would allow, and eventually China ceased its quota policy altogether. In 2015, China ended its practice of export restrictions and special taxes, allowing the market to fully stabilize.

Made in China 2025

In the aftermath of the Rare Earth Crisis, China aims to preserve its position as a price setter in the market, which will allow it to maintain a stable and inexpensive flow of REEs into the domestic economy for the production of advanced electronics. China is not just the dominant producer of REEs; it is also the dominant consumer. From the onset of its entry into the REE market, China has managed low-cost production of the refined metals through state support. As it emerged as the largest consumer of the metals, China has benefited from this competitive position. With the advent of the Made in China 2025 strategic blueprint, China's strategy for its future market position is clear: China feels it must maintain a stable and low-cost supply of REEs for its own advanced electronics manufacturing industries for the foreseeable future.

The Made in China 2025 blueprint was released in May 2015, as "the country's first ten-year action plan focusing on promoting manufacturing."[52] The plan advocates a strategy to transform China into a leading manufacturing power by 2049, with this stage designed to innovate in the manufacturing sector to emphasize advanced and "green" manufacturing sectors, as well as a shift toward developing Chinese brands. The plan focuses on ten "key sectors": new information

---

50. Johnston 2013.

51. Cox and Kynicky 2018.

52. State Council 2015, n.p.

technology, high-end automated machine tools, aerospace equipment, ocean engineering equipment, high-end rail transportation equipment, energy-saving and new energy automobiles, electrical equipment, farming machines, new materials, and high-end medical equipment and medicine. With the possible exception of farming machinery, all of these sectors will rely on steady, low-cost access to REEs. Moreover, the plan calls for improving techniques for processing important materials, such as the rare earths.

The consolidation of economic strategy that emerged as the Made in China 2025 blueprint issued by China's State Council in 2015 reveals the full value of controlling the REE market for the domestic economy. As Seaman writes,

> [T]his blueprint emphasizes nationalizing much of the value chain for new strategic industries—setting goals for Chinese production to account for up to 70 percent or 80 percent of the value of the domestic market for products such as new energy vehicles, new and renewable energy equipment and high-tech medical devices. In this respect, the export quotas, taxes and other price-distorting measures put in place until 2015 gave Chinese firms an added advantage of stable rare earth supplies at comparatively lower prices than consumers abroad, and even enticed some value-added production and technology transfer in downstream industries overseas to China.[53]

As such, the REE market is inextricably linked to China's broader manufacturing goals. Just as the core market for REEs is small but linked to huge portions of the global economy through supply chains, China's reliance on the steady supply of REEs is linked to its ambitious domestic economic goals to leap over the middle-income trap, whereby growing economies rapidly advance out of the lower-income stage but cannot make the final transition to high-income economies.[54]

Given its limited success during the Rare Earth Crisis, China has not placed a premium on using its market power as a source of political leverage. Recently, Chinese President Xi Jinping highlighted the importance of rare earths in global supply chains while touring a massive rare earths processing plant in Shanghai in May 2019.[55] In this case, however, the behavior was likely to signal credible deterrence against additional casualties in the trade war between the United States

---

53. Seaman 2019, 26–27.

54. World Bank 2013.

55. Hearty and Alam 2019; Bradsher 2019.

and China in 2019 and 2020. Regardless, events like these are exceedingly rare given the duration of China's market power. Looking at the long arc of Chinese dominance in the REE market and its plans for REEs moving forward, the more sensible position is that China seeks to maintain its status quo as the price setter in the market. In this sense, China's role in the REE market is similar to Saudi Arabia's role as the swing state in oil production or Russia's local role as the price setter for natural gas in Europe. The market power is real, but prices cannot be pushed too high without triggering significant adaptation by other players in the global market.

Market Power Motivation

A key component of China's strategy to maintain its dominant position in the REE market involves access to resources in the East and South China Seas. The seabed underneath these waters off the coast of China contain major deposits of REEs. The ability to engage in deep-sea mining to extract these resources would help China move toward its goal of having access to a stable supply of REEs and preserving its market share. However, China's access to these resources is hampered by long-standing property rights disputes. Given this, China's market power goals have motivated it to aggressively press its territorial and maritime claims in in the East and South China Seas.

East China Sea

In the initial decades of the Diaoyu/Senkaku dispute, the primary economic motivation for control over the islands was the surrounding deposits of hydrocarbons. The publication of reports by the United Nations in the late 1960s detailing the existence of significant oil and gas reserves in the East China Sea motivated countries to press their claims for sovereignty over the islands in the 1970s.[56] Competition over access to these hydrocarbons continued in the twenty-first century. As we described earlier, contention over China's drilling of natural gas near Japan's claimed EEZ boundary led to an escalation of the dispute between China and Japan in the early 2000s. This spike in tensions subsided with the eventual announcement of the joint development agreement in 2008.[57] Nevertheless, the two countries have yet to pursue joint hydrocarbon production in this area,

---

56. Shaw 1999.

57. Tsai 2016.

and China has continued to move forward with its unilateral efforts to drill for oil and gas.[58]

Needless to say, access to hydrocarbon resources remains one reason why China continues to press its claims in the East China Sea. In 2014, the US Energy Information Administration estimated 200 million barrels of proved and probable oil reserves and between 1 and 2 trillion cubic feet of proved and probable natural gas reserves in the East China Sea. Some Chinese sources claim there may be as many as 70–160 billion barrels of oil and as much as 250 trillion cubic feet of undiscovered natural gas, mostly in the Okinawa Trough.[59] However, in the past decade, another hard commodity has emerged as a key economic motivator for China's expansionist stance in the East China Sea: REEs.

Large deposits of REEs were discovered in the East China Sea, in part as a result of the explorations interested in hydrocarbons.[60] Their discovery occurred in advance of the ability to extract them, but as forward-thinking economies, it is unsurprising that both China and Japan would shape their policies in the East China Sea around the expectation of tapping into this set of resources. The discovery of hydrothermal activity in the Okinawa Trough dates back to 1984, and a detailed but preliminary discussion of the potential REE deposits emerged in a study in 2001.[61] One of the major hydrothermal fields in the Okinawa Trough lies at the southern end of the Trough, just south of the Diaoyu/Senkaku Islands.

The discovery of REE deposits in the East China Sea (along with the deposits in the South China Sea, discussed in the following subsection) provides a potential strategy for China to ensure a continued supply of REEs even if land-based sources become limited by diminishing supply or environmental demands. Globally, over the last thirty years, pressure to prioritize the environment has erected new barriers to continued REE mining on land. In 2001, Scott concluded that "much of this new marine activity is being driven by environmental and land-claim problems being increasingly encountered on land."[62] The combination of REE deposits, oil, and gas in the Okinawa Trough provide an economic basis for the enduring dispute between the two states over the contested islands. China has argued for years, and formally presented to the United Nations, that its UNCLOS delimited economic control in the East China Sea should be based

---

58. Kyodo 2019.

59. US Energy Information Association 2014.

60. Hongo et al. 2007; Xu et al. 2011.

61. Cao et al. 2018; Zhang, Hou, and Tang 2010.

62. Scott 2001, 92.

on the assertion that the Diaoyu Islands are part of China. Japan, unsurprisingly, has pushed back. As the market for REEs increased in importance for both states after 2010, China's motivation for expanded control in the East China Sea has been matched by Japan's motivation to delay China's dominance. With the recent discovery of a very large deposit of REEs within Japan's EEZ southeast of the Ogasawara Islands, China will need to maintain low market prices to prevent Japan from investing in the development of deep-sea mining technology.

South China Sea

China's market power goals with respect to REEs are not limited to the East China Sea and Japan. Its desire for control of the South China Sea is similarly motivated by this market and China's role as the dominant producer and consumer of rare earths. Control over the Spratly Islands and the Scarborough Shoal in the South China Sea would represent a large jump in economically valuable territory for China once we factor in the economic value of REEs and hydrocarbons in the surrounding EEZs offered by sovereignty over the islands. Large flows of dissolved REEs have been discovered in the South China Sea, along with polymetallic nodules and sulfides rich in REEs.[63] If China can gradually insert its physical presence and increase control over the Spratly Islands as well as elsewhere, it will bolster its claim to exclusive rights to harvest these commodities from the surrounding waters. Moreover, these enhanced and artificial islands provide a base from which to increase exploration efforts for REEs. With most of the seabed unmapped, particularly deep-sea areas, China is betting on the economic advantages of expanded South China Sea sovereignty.

Ultimately, such efforts could provide China with the stability it seeks in providing low-costs REEs to its domestic manufacturing interests. Whether firms under the Philippine or Chinese flag extract these resources, much of them will be consumed by the Chinese economy. The issue is not whether China will be able to obtain these resources, but rather the extent to which China will be able to prevent other states from developing the power to affect their prices in the Asian market. Perhaps more importantly, as we will explore in more detail in the next section, ISA mining code changes threaten to increase the price of REEs due to a new focus on environmental concerns with respect to mining deep seabeds for minerals. If China is able to absorb enough of this territory, it can avoid both potential threats to its current market power benefits.

---

63. Alibo and Nozaki 2000; Long 2020; Wheedon 2019.

Incompatible International Institutions

International institutions provide a readily available set of rules and procedures that China and its fellow disputants could use to resolve their disputes over the control of maritime resources in the East and South China Seas. However, China has largely avoided any serious attempt to pursue such an institutional solution. In particular, China has been unwilling to turn to legal dispute resolution under the auspices of UNCLOS. This reluctance is due to the incompatibility between China's market power goals and the design of UNCLOS and the related ISA. In both the East and South China Seas, the rules outlined by UNCLOS create uncertainty as to whether China's expansionist claims would be upheld by an international court or arbitration panel. At the same time, a renewed focus on environmental concerns within the ISA would pose a challenge to China's ability to extract rare earths at relatively low costs if it were to receive an unfavorable legal ruling on its claims in the South China Sea.

United Nations Convention on the Law of the Sea

The United Nations Convention on the Law of the Sea emerged in the 1990s as a powerful new tool to delineate sovereign versus international waters. UNCLOS was crafted in 1982 and only entered into force in 1994 as states ratified the treaty through their domestic institutions. By 1996, all of the disputants in the East and South China Seas (except Taiwan) had ratified UNCLOS.[64] This institution represented a departure from traditional approaches to resolving maritime disputes. Whereas in the past, states would bring their historical claims to bear in their negotiations, the UNCLOS regime is grounded in a distance-based, ahistorical set of rules. These rules provide equality to all states with respect to the distance from their shores that may constitute not only sovereign territory but also an EEZ.

Core to the functioning of UNCLOS is the definition of territorial seas versus international waters. All states bordering with seas shall consider the twelve nautical miles beyond their baseline to be sovereign territory.[65] With respect to this territorial sea, the Convention specifies:

---

64. As Taiwan is not a member of the United Nations, it cannot be a party to the Convention.

65. The normal definition of a "baseline" is the low-water point of the coastline, with numerous exceptions provided for reefs and islands.

1. The sovereignty of a coastal State extends, beyond its land territory and internal waters and, in the case of an archipelagic State, its archipelagic waters, to an adjacent belt of sea, described as the territorial sea.

2. This sovereignty extends to the air space over the territorial sea as well as to its bed and subsoil.

3. The sovereignty over the territorial sea is exercised subject to this Convention and to other rules of international law.[66]

Thus, within the territorial sea, the Convention assigns sovereignty to the state from top to bottom, including the airspace above the water, the water itself, and the seabed and subsoil. The UNCLOS code identifies the water beyond these territorial seas as international waters.[67]

In addition to the territorial sea, UNCLOS also identifies two additional areas within international waters in which states have exclusive access to marine resources. First, states enjoy an EEZ that stretches up to 200 nautical miles from the baseline. An EEZ is an area in international waters in which the state has a sovereign right to all economic resources below the surface of the sea. Of particular relevance to our discussion, a state has jurisdiction over the exploration and exploitation of any natural resources found in the seabed within its EEZ. Second, a state also has a sovereign right to explore and exploit resources within its continental shelf, which can extend beyond 200 nautical miles from the baseline. As defined by the Convention, the continental shelf "comprises the seabed and subsoil of the submarine areas that extend beyond its territorial sea throughout the natural prolongation of its land territory to the outer edge of the continental margin."[68] Thus, unlike an EEZ, which provides a right to all resources below the surface of the sea, the continental shelf only provides a right to seabed resources.

While the advent of the Law of the Sea has transformed the way the world and the global economy have organized around the convention, the design of UNCLOS is not well suited to resolve a maritime jurisdiction dispute in tight quarters, such as in the East China Sea. As Ramos-Mrosovsky writes, "In any sea less than 400 nautical miles across, areas of maritime jurisdiction will overlap. The East China Sea is only 360 nautical miles across at its widest point."[69] The rules laid out by UNCLOS, however, do not explicitly spell out how maritime

---

66. United Nations 1982, n.p.

67. Asariotis and Premti 2018; United Nations 1982.

68. United Nations 1982, n.p.

69. Ramos-Mrosovsky 2008, 911.

boundaries should be drawn when states have overlapping EEZ or continental shelf claims. The Convention only states that this delimitation "shall be effected by agreement on the basis of international law . . . in order to achieve an equitable solution."[70] If states are not able to resolve their differences bilaterally, UNCLOS does lay out a set of institutional procedures that states can follow to reach such an agreement, including international arbitration or adjudication. However, such legal dispute resolution carries too much uncertainty for both China and Japan.

Given the vagueness of the criteria for delimiting EEZs and continental shelves between states with opposing coasts, it is not surprising that China and Japan have proposed different interpretations of UNCLOS as it relates to the location of their maritime boundary in the East China Sea. Japan has proposed an EEZ boundary based upon the median line between the states' baselines. A median line is a common starting point for drawing maritime boundaries that meet the "equitable solution" requirement outlined by UNCLOS. China, on the other hand, claims that its EEZ extends further to the east and that it should have jurisdiction over the seabed in most of the East China Sea due to a continental shelf that extends to the Okinawa Trough.

In addition, any resolution of these maritime claims is conditional upon the resolution of the territorial dispute over the Diaoyu/Senkaku Islands. The state with sovereignty over this island chain would likely be entitled to a territorial sea of twelve nautical miles around the major islands. Moreover, these islands can potentially be used as base points for establishing the extent of the state's EEZ. If China were awarded sovereignty of over the islands, this would bolster its claim to a right to the continental shelf up to the Okinawa Trough. On the other hand, Japanese ownership of the islands would likely undermine this claim and benefit Japan's assertion of a maritime boundary farther to the west. As with the maritime dispute, international law is unclear as to which state should have sovereignty over the Diaoyu/Senkaku Islands.[71] Thus, the presence of a related territorial dispute adds further uncertainty to any potential legal ruling on the competing claims in the East China Sea. Given this, neither side is willing to risk the uncertainties associated with using international institutions to resolve the dispute.

Similarly, settling the disputed claim over the South China Sea through arbitration or adjudication would be a risky strategy for China. China's claim to sovereignty over the South China Sea within the nine-dash line is based, at least in part, upon "historic rights." Such a historically based claim is

70. United Nations 1982, n.p.

71. Ramos-Mrosovsky 2008.

inconsistent with the approach established by UNCLOS to determine areas of maritime jurisdiction.[72] Additionally, China has been strategically vague about its territorial and maritime claims in the South China Sea. In a note submitted in 2009 to the United Nations in response to maritime claims made by Vietnam and Malaysia, China stated that it "has indisputable sovereignty over the islands in the South China Sea and the adjacent waters, and enjoys sovereign rights and jurisdiction over the relevant waters as well as the seabed and subsoil thereof."[73] Attached to the note was a map of the South China Sea that included the nine-dash line. While this line has its roots in a map published in the late 1940s, China has never explicitly clarified whether it is meant to delimit China's maritime claim or to merely identify which islands China claims sovereignty over.[74]

The Permanent Court of Arbitration (PCA) case between the Philippines and China illustrates the challenge China faces in achieving its expansionist goals through a legal process. The PCA rejected China's historical claim to the nine-dash line. Moreover, it ruled that all of the Spratly Islands and the Scarborough Shoal are classified as either low-tide elevations or "rocks" under international law. Given this, none of these features generates an entitlement to an EEZ or continental sea. Therefore, even if it were determined that China has sovereignty over some of these islands, China would be limited in its ability to access seabed resources in that area of the South China Sea.[75]

Given this relatively weak legal claim, it is perhaps not surprising that China refused to participate in the proceedings at the PCA. Since 2016, China has rejected the legitimacy of the ruling and has continued to pursue its expansionist efforts in the South China Sea. For example, in spite of the fact that the PCA explicitly ruled that Mischief Reef is within the Philippines' EEZ, China has continued its efforts to build and fortify this artificial island. Since China is unable to achieve its expansionist goals in a legal forum, it has instead relied upon a combination of strategic delay and gray zone tactics. As long as China can resist the resolution of the property rights claims, it holds the upper hand in implementing its cabbage strategy of slowly expanding and consolidating its area of control in the South China Sea. Given its economic dependence upon China, the Philippine government cannot escalate the conflict outside of the courts, and

---

72. Dupuy and Dupuy 2013; Hayton 2018.

73. Permanent Mission of the People's Republic of China to the United Nations 2009, n.p.

74. Dupuy and Dupuy 2013.

75. Reed and Wong 2016.

this unsettled zone of disagreement favors the slow but inexorable march toward de facto Chinese territorial control.

International Seabed Authority

An important question remains, however. Why is sovereignty important to China in an economic sense? The answer may be found in the institutional evolution of the International Seabed Authority. The ISA's charge dates back to 1994 and the original UNCLOS agreement, but only recently has this governing body emerged as an important tool in navigating the mining of the deep sea. In particular, the ISA's chief mission is to regulate mining in the "Area," which is the term that UNLOS uses to refer to the seabed beyond the limits of national jurisdiction. The Area lies in international waters beyond any state's EEZ or continental shelf. The ISA is charged with providing mining rights to states and contractors to mine the Area, a legal structure for contracting, and a governing body for establishing and maintaining regulations. In 2020, the ISA mining code was still in draft form, with the possibility of adopting the draft in the near future.

One of the core activities that the ISA expects to manage is the extraction of polymetallic nodules. These nodules are small rock structures (up to 20 cm in diameter) that are generally half-buried in the seabed sediment, with the highest concentrations between 4,000 and 6,000 meters below sea level.[76] These nodules can be rich in cobalt, manganese, nickel, and rare earths, but until recently the mining of the nodules was prohibitively expensive in comparison to land-based mining. China is a world leader in the development in deep sea mining technology, however, and is poised to take advantage of this bountiful resource. In 2020, the Chinese firm, China Minmetals, signed a fifteen-year contract with the ISA to explore the Clarion-Clipper Zone in the Pacific Ocean for polymetallic nodules. China has signed similar contracts for exploration in the Southwest Indian Ridge and the West Pacific Ocean. Aside from the potential to mine the seabeds for minerals, these contracts all share something in common in that none of them falls within the nine-dash line. China has shown a willingness to work with the ISA and within its legal framework, but it seems that willingness may not extend to the seabed lying within the nine-dash line in the South China Sea.

The ISA's mission of working toward a mining code that allocates mining rights within the structure of the UNCLOS framework may be the reason for China's reluctance. Any mining that would take place in the Area would be governed by the ISA mining code. Moreover, contracts issued by the ISA must be

---

76. International Seabed Authority 2006.

approved by the ISA's Council, which is composed of thirty-six member states. Even though China has been an active member of the organization, relying on the ISA to allocate mining rights risks China's ability to assure a steady supply of rare earths into the Chinese economy over the next three decades.

A recently renewed focus on environmental stewardship may provide another source of risk for China. At the 26th annual session of the ISA in Jamaica early in 2020, environmental protection was a major theme of discussion.[77] In 2019 the ISA Assembly approved a strategic plan for the next five years that included strategic directions to preserve the marine environment and to share the economic benefits of mining the Area.[78] In all of these activities it is reasonable to assume that China is a willing partner. With these strategic directions in place and the authority of the Council, however, it is also reasonable to conclude that China prefers not to treat the zone of the South China Sea within the nine-dash line as being part of the Area. China cannot risk leaving access to these waters to the ISA within UNCLOS, and thus it is in pursuit of expansion to control its own destiny with respect to extracting these resources from the sea.

Economic Constraints

The design of international institutions helps to explain why China has been unwilling to pursue a legal settlement of the property rights disputes in the East and South China Seas. However, China has been similarly reluctant to use military force to expand its control over resources in these disputed seas. This restraint is due in part to the nature of China's international economic relationships. In both the East and South China Seas, economic interdependence constrains China from escalating these disputes too far. However, the nature of these economic constraints varies between the two cases. In the East China Sea, China's dependence on Japan for trade and investment has required China to exercise restraint in its expansionist activities. On the other hand, China enjoys an asymmetrically interdependent relationship with its South China Sea neighbors, which has allowed China to take more aggressive steps to the south. However, China's overall dependence upon the global REE market constrains China from turning to use of force capture control of the South China Sea. These economic constraints help to explain both China's decision to pursue strategic delay, as well

---

77. Kantai et al. 2020.

78. International Seabed Authority 2019.

as the divergence in the tactics it has used to pursue its expansionist goals in the two sets of disputes.

Interdependence between China and Japan

As we have seen, China has largely refrained from using its military might to press its expansionist claims in the East China Sea. Even when China has taken more combative steps, like deploying naval ships to disputed waters, it has steered clear of escalating the dispute too far. This restraint on China's part is largely due to the economic constraints that it faces. In particular, the high level of economic interdependence between China and Japan prevents China from pursuing a more aggressive strategy. While the economic incentives associated with the islands are critical to China's economy, so is its economic relationship with Japan. Exiting that relationship would be very costly to China, so the PRC has strong incentives to avoid taking actions that would risk a disruption of its trade and investment ties with Japan.

The nature of the Sino-Japanese economic relationship has evolved over time. In the initial decades of the Diaoyu/Senkaku dispute—from the 1970s to the 1990s—the economic interdependence between China and Japan was largely asymmetrical in Japan's favor. Koo notes that in bilateral trade between the countries in the 1970s, "China depended on Japan for trade far more than Japan did on China."[79] This asymmetric nature of their trade relationship was maintained as China's international trade increased in the wake of its economic reforms under Deng Xiaoping. Through the 1980s, China's trade dependence on Japan continued to increase at a faster rate than Japan's dependence on China.[80] Due to these economic constraints, China set aside escalating its territorial claims over the islands in favor of fostering economic ties with Japan. The asymmetry in the economic interdependence thus allowed Japan to rebuff China's territorial claims without fear of militarized violence.

In the twenty-first century, the economic interdependence between China and Japan has become more symmetrical as Japan has become more dependent upon trade with China. From 1997 to 2006, Japan's trade with China as a share of its GDP nearly tripled.[81] Currently, both China and Japan rely heavily upon one another for exports. Japan's top export target is China, while Japan is China's

---

79. Koo 2009, 218–219.

80. Koo 2009.

81. Ibid.

second-largest export target. In 2018, 19.5 percent of Japanese exports went to China, while about 6 percent of Chinese exports headed to Japan. Similarly, China is Japan's largest source of imports, while Japan is China's second-largest import partner.[82] Given this interdependence, neither party can afford to sever economic ties. Thus, both states can safely rattle their sabers without fear of escalation.

China's economic dependence upon Japan extends beyond the trade arena. By the mid-1990s, Japan had emerged as the largest source of development and investment in the Chinese economy, including $3.2 billion (US) in loans and investments from Japan to China in 1995 alone.[83] China continues to enjoy large inflows of foreign direct investment (FDI) from Japan. In terms of the use of FDI inflows to help develop manufacturing infrastructure, Japan is perhaps the most important partner for China. If China decides to become more aggressive in its pursuit of the islands, Japanese foreign investors may interpret this move as a sign that their capital is no longer safe in the mainland, and they may shift investments elsewhere. Latorre and Hosoe simulate a negative shock to Japanese FDI in China and determine that it would reduce production (and exports) from Japanese affiliates in China, depreciate the renminbi, and particularly hurt China's service sector.[84] Thus, the current economic constraints are sufficient to prevent China from escalating the dispute militarily and limit China's ability to press Japan to resolve the territorial dispute.

China's economic dependence has allowed Japan to maintain a policy of strategic delay in the East China Sea. Given the costs of economic exit, China is not willing to use military force to seize control of the disputed islands or press its maritime claims. Instead, the mutual interdependence that characterizes more recent years has enabled both states to safely voice their political demands without fear of escalation. Thus, it is of little surprise that China continues to take provocative steps like announcing an air defense identification zone that includes much of the East China Sea and conducting hundreds of incursions by government ships into the territorial waters around the Diaoyu/Senkaku Islands. At the same time, Japan has responded by increasing its posture to defend its territorial claim by significantly increasing its coast guard presence in the area.[85]

---

82. UN Comtrade data accessed from World Bank 2020. The United States is China's top export target, while South Korea is China's largest source of imports. Technically, Hong Kong is the second largest export target for China, but we exclude it for this discussion.

83. Koo 2009, 225.

84. Latorre and Hosoe 2016.

85. Burke et al. 2018.

Asymmetrical Interdependence in the South China Sea

Compared to the dispute in the East China Sea, China has been more willing to take aggressive steps to press its claims in the South China Sea. By constructing artificial islands and building up its military presence, China uses its cabbage strategy to expand and consolidate its area of control in the South China Sea. This more assertive stance is in part a function of the asymmetrical economic interdependence that China enjoys with its South China Sea neighbors. For example, the Philippines, Vietnam, and Malaysia would face much larger costs from exiting their economic relationships with China than China would face from severing ties with any one of these countries. Given this, these countries are not able to constrain China in the same way that Japan can in the East China Sea. To illustrate this asymmetric interdependence, we focus on the economic relationship between China and the Philippines, but a similar story would hold if one were to look at China's relations with other South China Sea claimants.

The economic relationship with China is very important to the Philippines. China is one of the Philippines' top trading partners. In 2018, 12.9 percent of Philippine exports went to China, and 19.6 percent of its imports came from China.[86] Only the United States, Japan, and Hong Kong received a larger share of Philippine exports than China. China is the Philippines' largest source of imports, providing almost twice as much as the next country, South Korea. In addition to this trade relationship, the Philippines also has begun to depend more on FDI from China. FDI flows from China to the Philippines have increased significantly under the Duterte regime. China was the Philippines' largest source of FDI in 2018 and its second largest provider of FDI (behind Singapore) in 2019.[87]

In contrast, China is much less dependent upon the Philippines. While the Philippines is an important trade partner for China as well, it falls below the top ten trading partners. The Philippines accounts for just 1.4 percent of China's exports and 1 percent of China's imports.[88] Additionally, unlike Japan, the Philippines does not provide a significant source of FDI for China. Thus, the nature of the trade and investment relationship between China and Philippines indicates an asymmetric economic interdependence. Given the volume of trade and investment that it would lose, Philippines would suffer significant exit costs if it were to sever its economic ties with China. China, on the other hand, would find economic exit less costly.

---

86. UN Comtrade data, accessed from World Bank 2020.

87. Philippine Statistics Authority 2020.

88. UN Comtrade data, accessed from World Bank 2020.

This asymmetrical economic interdependence indicates that the Philippines cannot credibly escalate its territorial dispute with China over the Spratly Islands regardless of military concerns. Since the Philippines cannot afford to exit its economic ties with China, economic interdependence constrains the Philippines. Thus, in the current economic environment, China can expect the Philippines to make political and legal demands without any real fear of escalation to militarized violence. On the other hand, since China could more easily cut its economic ties with the Philippines, this economic relationship does not prevent China from taking a more aggressive stance in pressing its territorial and maritime claims in the South China Sea.

This interdependence, however, can be misleading if one concludes that China is unconstrained economically. While China enjoys asymmetrical power in its dyadic economic ties with South China Sea neighbors, it is still constrained by the potential adaptation of the global REE market if it begins to use its REE market power as a tool of economic statecraft or if it renews its attempts to extract rents from importers of REE materials through quotas or taxes. While China is the dominant producer of REEs and processed oxides, it has never controlled the majority of global reserves. Rare earths are not that rare. They are, however, expensive to extract in that they require significant infrastructure for mining. Thus, China's market power is constrained because states like the United States and Japan could rapidly invest in their mining capacity should China push prices too high and availability too low.

The result is a level of economic interdependence that acts as a constraint against the use of military force. The use of force would cement the consideration of REE supplies as national security issues for Japan and for the United States, thereby enabling costly investments into restarting land-based mining as well as new ventures in deep sea mining.[89] As such, China's constraints look similar to Russia's market constraints with respect to exporting natural gas.[90] Like Russia, China cannot push prices of REE materials too high, or it will stimulate new entrants to the market.

In order to maintain its subsidies of providing low-cost rare earth materials to its domestic economy, China needs to sell a significant quantity of rare earth materials abroad. Its dominance in the market allows for this strategy, but that would be compromised if other firms and state-controlled interests enter the market and provide alternative supplies to buyers. China's top priority (and challenge) is therefore to maintain a stable swing position in the market, rather

89. Bilsborough 2012; Butler 2014.

90. See Chapter 6.

than monopolizing and then exploiting its power through quotas and price hikes. Prices need to stay high enough to sustain the industry and preserve China's competitiveness in manufacturing next-generation electronics, but low enough to keep others from investing in their own mining capacities. The result is an impetus to expand sovereign control over the seas that can provide a steady supply of REEs to extract for the next few decades.

Stepping back, then, we can see important differences in China's dyadic interdependencies with Japan compared to its South China Sea neighbors. Japan's economic influence on the Chinese economy gives it the ability to push back against the expansionism that would further entrench China's market power. The Philippines and the other South China Sea claimants, on the other hand, cannot afford to confront China outside of the arena of international institutions. As such, it is unsurprising that China can be more assertive in its tactics in the South China Sea. At the same time, however, China faces economic constraints due to the abundance of REEs worldwide. As long as China avoids the kind of price spikes that resulted from its actions in 2010, it should be able to maintain its ability to deter new entrants to the REE market.

Alternative Explanations

Here we examine two alternative explanations for China's behavior in the East and South China Seas: one based on China's reputation vis-à-vis domestic and international property rights disputes, and one based on power and the pursuit of power. Both are important lines of argument, and when examining the full history of events, it would be improper to ignore these motivations. The explanatory power of the market power politics argument, however, provides important insights even in the face of these alternatives.

Let us first examine the alternative explanation of reputation dynamics. Perhaps the most obvious alternative explanation of China's behavior in the South China Seas is that China is simply following a reputation-driven strategy. China may view any individual island dispute with the concern that it would set a precedent for other territorial disputes, both maritime and land based.[91] Moreover, China may worry that any legal resolutions reached with neighboring states may lead to new claims internally. With intensifying concerns and conflict with western provinces, China may be loath to give credence to legal resolutions that unravel its grasp in Xinjiang and Tibet.

---

91. Melin and Grigorescu 2014; Walter 2009; Zhang 2019.

If we focus the reputation frame only on maritime claims, China's disputes with Japan share the same characteristics as this dispute with the Philippines. The problem with this explanation is that China has been too lenient with Japan's claims over the last half-century. If China was simply pursuing a reputation strategy, then it would be forced to be much more aggressive in trying to reclaim the Diaoyu/Senkaku Islands. The reputation story may lead to the additional prediction that China will stall until it has reached its desired bargaining position with respect to all of its maritime disputes, but the basic logic of our analysis still holds. If, however, one assumes that China is concerned that resolving sovereignty issues for the relatively uninhabited Spratly Islands, for example, would trigger a cascade of renewed pressure to grant territory to separatists in Xinjiang and elsewhere, then the reputation explanation would dominate. In our view, this assumption is tenuous because of the remarkable difference in populations, and China's willingness to tolerate the unfavorable status quo with Japan counters the strict reputation argument.

One could also argue that we have ignored other strategic motivations, including power dimensions. A central explanation for China's lack of aggression against Japan for most of this time period is Japan's dominant navy. Additional concerns include the importance of the United States in the region, its long-standing alliance with Japan, and the lingering geopolitics of the Cold War. This source of influence is likely to be most relevant in the first half of this temporal analysis until the end of the Cold War, but clearly the US-Japan defensive alliance continues to exist at full strength. There is less variance along this dimension than would be required to explain China's newfound assertiveness with a surge in military power. Moreover, the presidential transition in the United States from President Obama to President Trump in 2017 should have triggered major changes in China's strategy along the Pacific Rim. Whereas President Obama engaged in a pivot toward Asia and devoted time and resources to countering China's rise, President Trump has unraveled that focus. The dramatic departure of the United States from the region has not translated into a surge of more aggressive behavior on the part of China. Instead, the status quo has largely prevailed. As such, we conclude that power dynamics alone cannot fully explain China's continued use of strategic delay and gray zone tactics.

Conclusion

Ever since the discovery of rare earth resources surrounding the features in the East and South China Seas, there has been pressure for China to seek to develop these resources in search of a market power opportunity. As China becomes more economically powerful, Japan will continue to try to prevent China from establishing more market power. Less economically powerful states, such as the Philippines and Vietnam, do not have that option and are likely to remain frustrated with China's unwillingness to engage in dispute resolutions driven by UNCLOS. The extent to which this market power is valued is in part a function of stable market prices. The recent move in the ISA toward prioritizing the preservation of the environment can be seen as a threat to China's low-cost production strategy. As long as China perceives the need to provide a steady supply of affordable rare earths for domestic production, it will continue to resist ISA governance in the South China Sea.

The 2010 trawler incident, however, stands out as a clear sign to China that it must exercise great caution when using its rare earths market power as a tool of economic statecraft. China has the ability to leverage its edge in the South China Sea to expand its military presence and literally build an argument for sovereignty, but it learned in 2010 to use care when restricting the supply of refined rare earth minerals to the global economy. Much as Russia faces the constraint of high natural gas prices pushing Europe toward a more aggressive adoption of green energy sources, China cannot leverage rents in this market or withhold its exports for political reasons without spurring new investment and entry into the market.

Based on this market constraint, we do not anticipate that China will use its market power as a source of economic statecraft in geopolitics, despite veiled mentions of the possibility by the Chinese government in 2019. Unless the likelihood of new entrants into the market drops precipitously, China will not risk a price shock that triggers entry. At the same time, there is no reason to conclude that China will stop its inexorable march to occupy more islands in the South China Sea. The economic constraints on China's neighbors confine them to institutional paths of dispute resolution, which China has ignored in this arena. Attempts to resolve these disputes will flounder if they do not address the underlying (or in this case, underwater) motivation that China must maintain its dominance of the REE market for the foreseeable future.

From Iraq and the United States and the Persian Gulf War to Russian expansionism and gray zone tactics and finally China's maritime push in pursuit of continued REE market dominance, it seems evident that market power politics can trigger war when important constraints are absent, and it can lead to lower levels of conflict when economic or institutional constraints make war too costly. With these exploratory cases in mind, in our final chapter we synthesize the lessons of the book and briefly examine a case where a set of disputes driven by market power opportunities ultimately led to a peaceful resolution.

8

Conclusion

OVER THE PAST seven chapters, we have shown how competition over market power can have very consequential effects on international relations. In normal economic exchange environments, buyers and sellers are price takers, and politics in the global political economy typically revolves around the coordination and manipulation of the effects of market prices. To promote efficiency in their economic relations, states create institutions that define property rights and facilitate dispute resolution. However, when states have opportunities to change market structures in order to provide their firms with the ability to set prices, normal economic exchange can break down and a more competitive environment can emerge.

Such market power competition, particularly in hard commodity markets, can create or exacerbate property rights disputes between states. Given the economic rents and political leverage that can accompany the ability to set prices in these markets, states may be motivated to take aggressive action to expand their territorial reach. This market power motivation can sometimes lead to war. However, when states are economically interdependent, they may be constrained from turning to violence to achieve their market power goals. This can open up an opportunity for institutional settlements. However, in some cases, institutional rules and procedures can preclude states from reaching a settlement in line with their market power ambitions. When this happens, states may prefer to strategically delay a dispute settlement, which can provide an opportunity to gradually accumulate market power over time through salami tactics.

To explore how these strategic dynamics play out empirically, we examined cases of market power competition in three hard commodity markets. First, we saw how the desire to set prices in global oil markets motivated Iraq to invade neighboring Kuwait and then set its sights on Saudi Arabia. To prevent such a significant shift in oil market, the United States led a military coalition to block

Iraq's advances and push Iraqi forces out of Kuwait. Next, we turned our attention to Russia's ongoing efforts to maintain its market power in the European natural gas market. As part of its overall strategy to block potential competitors and secure its control over the transit of gas to European consumers, Russia has perpetuated territorial disputes with neighboring Georgia and Ukraine. Finally, we examined how China's desire to remain the predominant global supplier of rare earth elements underlies its ongoing expansionist efforts in the South and East China Seas. Given the incompatibility between institutional rules and its desired market outcome, China has avoided settling these maritime disputes and has instead relied on a strategy of delay and salami tactics.

This concluding chapter aims to tie up our discussion of market power politics in two general ways. First, we touch on some aspects of market power politics that we were not able to fully address in this book. In the next section, we provide a brief discussion of the 2018 settlement of the long-running dispute over the Caspian Sea. This allows us to consider some of the factors that might lead a state with market power motivations to abandon a strategy of delay and settle a property rights dispute. We then outline a set of questions for future research on market power politics. These include new theoretical questions that arose in light of the patterns we found in our case studies, as well as questions about how our model could be expanded beyond the set of countries and commodity markets that we have examined here.

Second, we reflect upon how our research informs our understanding of international relations. We highlight two important lessons of this book concerning the effects of market structure on conflict behavior and the limitations of international institutions. We then contemplate the future role of gray zone tactics by countries like Russia and China in light of our analysis of market power politics. Finally, we conclude the book with a discussion of some of the policy implications that follow from our research.

Moving from Strategic Delay to Dispute Settlement

The empirical cases analyzed in the previous chapters focused on strategies of war and delay in property rights disputes. Since these strategies represent the largest deviations from the expected behavior of states in normal economic exchange, this allowed us to better see how market power motivations can drive states to pursue expansionist, non-cooperative policies. This helps to explain the escalation to military force in the First Gulf War, as well as the gray zone tactics of Russia and China in recent years. However, this approach did not allow us to consider cases of successful settlement of property rights disputes in the face of market power motivations. This leaves open the empirical question of the conditions under which states would be willing to abandon a strategy of delay and agree to settle a property rights dispute. While a full-scale analysis of this question must be left to future research, here we briefly consider Russia's decision to sign an agreement to resolve the long-standing dispute over the legal status of the hydrocarbon-rich Caspian Sea.

Caspian Sea Dispute

The Caspian Sea is the largest inland body of water in the world and the location of significant hydrocarbon resources. The US Energy Information Agency estimates that the Caspian basin includes 48 billion barrels of oil and 292 trillion cubic feet of natural gas in proven and probable reserves.[1] Approximately 35–40 percent of these oil and gas resources are offshore. Until recently, the five surrounding states—Russia, Azerbaijan, Kazakhstan, Turkmenistan, and Iran— disputed the legal status of the Caspian Sea. This created uncertainty about which states had the right to access offshore hydrocarbon resources and construct pipelines across the sea. Since the breakup of the Soviet Union, Russia largely pursued a policy of strategic delay in negotiations over the Caspian Sea. However, in 2018, Russia finally agreed to settle the dispute, and all five littoral states signed the Convention on the Legal Status of the Caspian Sea. What led to this shift in strategy?

At the heart of the maritime dispute was the technical question of whether the Caspian Sea is a sea or a lake under international law. The unique features of this body of water makes its status uncertain. While the Caspian Sea is called a "sea" and contains salty water, it has no saltwater connection to the open seas. The question of whether it is a sea or a lake has a direct bearing on the ability of states to access the Caspian Sea and its resources. If the Caspian is a sea, it would fall under the jurisdiction of the United Nations Convention on the Law of the Seas (UNCLOS). In that case, control of the Caspian's resources would legally be divided under the rules of UNCLOS. The land-locked littoral states could potentially gain the right to use Russian waterways to access the open seas, and non-littoral states could potentially access the Caspian waters. On the other hand, if the Caspian is a lake, it would not be under the jurisdiction of UNCLOS. Its waters and resources would be divided up by the littoral states and would not be accessible to outside states.[2]

---

1. US Energy Information Association 2013a.

2. Boban and Lončar 2016; Zimnitskaya and von Geldern 2011.

Disagreement over the legal status of the Caspian Sea arose in 1990s. Iran and the Soviet Union had signed a pair of treaties in 1921 and 1940 that treated the Caspian Sea as a lake. These agreements gave both states equal access to the waters, which were closed to outside states. After the breakup of the Soviet Union, Iran largely wanted to continue following this principle, while the post-Soviet states of Azerbaijan and Kazakhstan preferred a solution following the rules of UNCLOS. Russia carved out a position in between these two extremes, which was commonly referred to as "common waters, divided bottom." Like Iran, Russia argued that the Caspian was a lake whose waters should be open to all five littoral states but closed to other states. On the other hand, Russia was willing to divide up the seabed resources according to the rules of UNCLOS.[3]

For over two decades, Russia largely pursued a strategy of delay. In a series of bilateral deals from 1998 to 2001, Russia reached agreements with Azerbaijan and Kazakhstan outlining their mutual maritime boundaries demarcating the parts of the seabed under their control. However, Russia insisted that any agreement outlining the legal status of the Caspian Sea must be agreed to by all five littoral states. A series of Caspian Summits from 1996 to 2014 were unsuccessful in resolving the dispute.[4] Russia had little incentive to push for an agreement, as it largely benefited from the status quo. Since its maritime boundaries were demarcated, Russia was able to access hydrocarbon resources in its area of control. Moreover, Russia had no reason to turn to international institutions to resolve the dispute, as this would create the possibility of an unfavorable ruling that could potentially allow outside countries to gain access to the Caspian.

Russia's strategic delay in the Caspian was part of its overall strategy to protect its market power position in the natural gas market. Continued legal uncertainty would help Russia maintain its monopsonist position as a primary buyer of Central Asian hydrocarbons and block additional competition in the European gas market. As we discussed in Chapter 6, in the initial post-Soviet era, all gas pipelines from Central Asia connected to Russia. Turkmenistan has long hoped to construct the Trans-Caspian pipeline, which would link Turkmenistan to Azerbaijan and allow it to export gas directly to Europe through the Southern Gas Corridor. However, the uncertain legal status of the Caspian Sea had been a major hurdle from keeping this project from moving forward. Thus, by delaying settlement of the dispute, Russia was able to prevent the construction of a

---

3. Zimnitskaya and von Geldern 2011.

4. Boban and Lončar 2016.

Trans-Caspian pipeline and keep out an additional direct competitor in the European gas market.[5]

At the Fifth Caspian summit, the five littoral states finally were able to reach an agreement on how to resolve the Caspian dispute. The Convention on the Legal Status of the Caspian, signed on August 12, 2018, largely followed Russia's "common waters, divided bottom" approach.[6] The agreement largely treats the surface as international waters and divides the seabed into territorial zones. Ships from all five littoral states will be able to navigate the waters, while military ships from non-littoral states will be prohibited. The convention also lays out the rules by which the seabed should be divided. States are free to build pipelines through their own territorial zones, but all five states would need to be consulted on the environmental risks of any pipeline project.[7]

Why Settle?

The Caspian Sea case illustrates some of the conditions when states may be willing to settle property rights disputes in the face of market power motivations. First, as we discussed in Chapter 3, states will be better able to reach a peaceful dispute settlement when international institutions are compatible with their market power goals. Unlike in the territorial disputes with Ukraine and Georgia, Russia was able to achieve its preferred legal settlement to the Caspian Sea dispute within the rules of international law. Given the ambiguous features of the Caspian Sea, Russia and the other littoral states could reach a settlement that did not need to abide by the rules of UNCLOS. This gave them additional flexibility in crafting an agreement. Moreover, by reaching a settlement outside of UNCLOS, Russia was able to guarantee that it would not be faced with unwanted interference by an international tribunal in the future.

While the compatibility of international institutions explains why a settlement may have been possible in the Caspian Sea but not in Russia's disputes with Georgia and Ukraine, it cannot explain the timing of Russia's decision to abandon a strategy of delay for dispute settlement. A state will shift from strategic delay to settlement when the benefits of settling the property rights dispute outweigh its ability to pursue its market power goals through strategic delay. The Caspian Sea settlement provides several important benefits to Russia. For one,

---

5. Grigas 2017, 231.

6. The convention needs to be ratified by all five signatory states before it goes into effect. As of June 2020, all signatories except Iran had ratified the treaty.

7. Kramer 2018b; President of Russia 2018.

the convention largely implements Russia's preferred "common waters, divided bottom" approach and establishes the legal status of the Caspian Sea outside the jurisdiction of UNLCOS. Additionally, to protect its regional security interests, Russia can guarantee its ability to deploy military ships throughout the Caspian Sea and prevent outside military forces from accessing the waters. Moreover, the agreement provides Russia with an opportunity to increase its level of cooperation with Central Asian states that are increasing their trade ties to China through the One Belt, One Road policy.[8]

At the same time, Russia's utility for strategically delaying a settlement of the Caspian Sea dispute was declining. To see this, one needs to consider why Russia chose to pursue strategic delay in the first place. Strategic delay allows a state to pursue various market power goals. For one, it can gradually accumulate power through salami tactics. Second, it can wait for a potential future change in its bargaining position to achieve a more preferred settlement. Third, it can perpetuate uncertainty about the property rights to gain or maintain an advantage over economic competitors.

Russia's goal in delaying a settlement on the Caspian Sea was to maintain legal uncertainty to prevent the construction of pipelines that would increase export competition in the European gas market. Unlike the territorial disputes in Crimea and Ukraine or the island disputes in the South China Sea, the maritime dispute in the Caspian Sea did not provide much of an opportunity to engage in salami tactics. Additionally, by 2018, Russia no longer had a strong incentive to delay in anticipation of achieving a better bargaining position. As the signed agreement indicates, Russia was already in a position to achieve its preferred "common waters, divided bottom" outcome.

Moreover, Russia's leverage over the other littoral states was at its peak and was likely to decrease in the future. Azerbaijan was moving toward completion of the final pipelines in the Southern Gas Corridor to Europe. Russia had also lost its monopsonist power in Central Asia as Turkmenistan shifted its gas exports to China. Additionally, increased Chinese economic presence in Central Asia threatens Russia's historical hegemony in the region. Given these circumstances, Russia had an incentive to lock in its preferred outcome to the Caspian Sea dispute and instead look for alternative routes to block a potential Trans-Caspian pipeline.

In the end, Russia may be able to achieve the strategic advantages provided by the Caspian Sea convention without sacrificing its privileged position in the European gas market. Russia's concessions in the Caspian Sea agreement that

8. Kramer 2018a.

open up the possibility of a future Trans-Caspian pipeline may increase European acceptance of the Nord Stream 2 pipeline, which will reduce Russia's reliance on Ukraine as a transit state. Moreover, while the convention reduces some of the legal hurdles to constructing the Trans-Caspian pipeline, it does not guarantee that a connector between Turkmenistan and the Europe will be built. Under the agreement, Russia has the ability to consult on the environmental impact on any pipeline project in the Caspian Sea. As Igor Brachikov, a Russian foreign ministry special envoy, noted, this consultation can "go on however long it takes."[9] Thus, despite the apparent resolution of the property rights dispute in the Caspian Sea, this veto power may still allow Russia to continue to use delay tactics to prevent increased competition in the European gas market. While the Caspian Sea dispute provides just one example of dispute settlement in the face of market power motivations, the conditions that led Russia to abandon its policy of strategic delay and agree to settle the dispute are largely consistent with the expectations of our model of market power politics.

Further Exploring Market Power Politics

The empirical cases that we explored in this book served as plausibility probes for our theory of market power politics. They allowed us to examine how the causal mechanisms that we outlined in our theory have played out in a few high-profile cases. In line with our theoretical expectations, we have seen how the desire to capture and maintain market power in key commodity markets has motivated the expansionist strategies of Iraq, Russia, and China. At the same time, variation in economic and institutional constraints have shaped their choice to turn to the use of force or rely upon strategic delay and gray zone tactics to achieve their market power goals. This gives us confidence that our theory helps to explain some aspects of international relations.

In some ways, however, these cases have just scratched the surface of our understanding of market power politics. Several open questions remain as to how these processes play out. Some of these questions arise in response to the empirical patterns that we found in our case studies. Other questions result from the fact that we focused our analyses on a particular set of states competing for market power in a small number of hard commodity markets. We briefly highlight four of these open questions here. While any attempt to answer them must be left to future research, this can perhaps be seen as a road map for future exploration of market power politics.

---

9. Ibid.

A Portfolio of Market Power Strategies

One question that arises in light of our case studies is how territorial expansion strategies fit within the portfolio of strategies that states can use to pursue their market power goals. In each of the cases that we explored, states pursued a variety of tactics, many of which did not involve expanding their territorial reach or exacerbating property rights disputes. In some situations, these included alternative strategies to gain or consolidate market power. For example, Russia and Gazprom have tried to vertically integrate the transportation of natural gas to Europe by gaining ownership of gas pipelines in transit countries and building bypass pipelines like Nord Stream that more directly connect to European customers. At other times, states have pursued strategies aimed at securing rents from commodity exports that would not necessarily provide price-setting capabilities. We saw an example of this when Iraq attempted to lobby Kuwait and the United Arab Emirates to reduce their oil output through the institutional channels of OPEC.

A common thread in the empirical cases is that territorial expansion was typically not a market power strategy of first resort. Instead, the states tended to turn to expansionist strategies when they were not able to achieve their goals through other means. Saddam Hussein only decided to invade Kuwait and seize its oil resources after he was unable to use institutional and diplomatic means to convince Kuwait to reduce its oil production in line with OPEC quotas. Similarly, Russia's expansionist efforts have centered around Ukraine and Georgia, two countries in which Gazprom's attempts to gain control of gas pipelines have failed. On the other hand, given Gazprom's success in purchasing the Belarusian gas transit company, Beltransgaz, Russia has not needed to take such an aggressive tack toward Belarus. Similarly, China has only turned to the idea of extracting rare earths from the seabed as it begins to face domestic pressure for the environmental effects of mining rare earths on land. The same environmental dynamic that led the United States to abandon its own rare earths mining capacity in favor of Chinese imports will push Chinese mining firms offshore and into international waters.

As outlined in Chapter 3, our theoretical model did not explicitly consider the choice between territorial and non-territorial market power strategies. Instead, we focused on factors that would potentially motivate states to take aggressive actions. However, given the costs and risks involved with the use of military force or gray zone tactics, it is not surprising that states would initially attempt other avenues of achieving their goals before turning to a strategy of territorial expansion. Thus, we expect that the aggressive strategies taken by the states in our case studies may be most likely in situations where alternative tactics are not sufficient to meet states' market power goals. An expanded version of our theoretical model that considers territorial aggression as part of a broader portfolio of strategies might help us better understand the timing and combination of tactics that states use to compete for market power.

Seeking versus Preserving Market Power

Another question that arises from our case studies involves the potential differences between the strategies that states use to gain market power and those that they use to maintain it. In the development of our theoretical model, we explained that market power competition arises between states that want to change the status quo and those that want to maintain it. Which side of this dichotomy a state falls on depends upon its current position in the market. When a state wants to gain price-setting capabilities for its firms, it is in a position of trying to change the status quo. On the other hand, after a state has gained market power for its firms and wants to preserve it, it moves into a position of taking steps to maintain the current market structure.

As we developed our theoretical expectations about the use of violence and strategic delay to pursue market power, we did not explicitly take into account the state's current position in the market. However, our empirical cases indicate that the nature of the territorial expansion strategies that states use may vary depending upon whether they are trying to gain or maintain market power. For its firms to have price-setting capabilities in a hard commodity market, a state typically needs to have access to significant reserves of the commodity. Thus, states aiming to gain market power in these markets generally need to shift the distribution of resources in their favor. For this reason, we expect that these states will mainly pursue expansionist strategies to gain control of strategic territory that gives them access to additional commodity resources. We saw an example of this when Iraq invaded Kuwait in order to augment its share of global oil reserves.

In contrast, states that are trying to preserve their market power may have a broader range of territorial strategies available to them. To be sure, these states will still have incentives to pursue territorial expansion to obtain additional resources for themselves. China's ongoing efforts to guarantee access to rare earth elements in the South China Sea provide an example of this. However, since their main goal is to maintain the status quo, states aiming to preserve their market power may also use these strategies to neutralize threats to their market position. As we have seen, Russia used territorial expansion to achieve and maintain its market power goals. For example, Russia has a recent history of destabilizing Georgia, which is a main transit state for a potential rival in the European gas market. As long as Georgia provides an alternative for Europe, Russia will be motivated to destabilize its neighbor.

On the other hand, Russia's expansionist activities in Ukraine have been partially aimed at increasing Russian leverage over its own transit state. The variation in these cases illustrates the important role that the status quo structure of the market plays in the strategic choices of states. Our expectation is that further exploration of the different strategies that states use to seek, preserve, or prevent market power would be a fruitful topic for future research. Japan's ability to inhibit China's expansion in the East China Sea, for example, has been nuanced in comparison to some of the other cases explored in this book.

Market Power Competition in Other Markets

A third question concerns the types of markets that fall within the scope of our argument. Our empirical cases examined the politics within three hard commodity markets: oil, natural gas, and rare earth elements. We chose to focus our plausibility probes on these markets for several reasons. First, the structure of these commodity markets can provide actors with opportunities to gain price-setting capabilities. In addition, states can have the opportunity to increase the market power of their firms in these markets through territorial expansion. Finally, these resources are highly salient to both consumers and producers. As we have seen, states have been willing to take aggressive steps to compete over market power in these particular commodity markets. However, it remains an open question as to whether competitive dynamics in other markets could lead to similar outcomes of war or strategic delay.

Unfortunately, there is little reason to expect states to avoid market power opportunities when they see them. Liberia's support for the Revolutionary United Front's incursions into Sierra Leone in 1991, for example, was influenced not only by the need to extract diamonds from its neighbor to the north but also by the desire to prevent Sierra Leone from emerging as an economic rival. Warlords in Sierra Leone associated cooperation and centralized governance with a loss of economic power and stymied peace in pursuit of local market power. In a more recent and nascent market example, not all states are interested in cooperating as the world races toward the discovery of treatments and vaccines for the COVID-19 virus. Early in its response to the crisis, for example, the United States government sought to procure German firm CureVac (which also has offices in Boston, MA) in an attempt to gain exclusive access to the vaccine that the firm was working toward. Looking backward, we can also see the importance of market power as a driving force in the intense violence between Native American tribes during the seventeenth century's Beaver Wars.

Lastly, it is worth pointing out that even advanced and complex economies such as the United States and Japan are not immune to the influence of market power politics. After all, the US decision to push Iraq's forces out of Kuwait and defend the Saudi Arabian border can only be explained by its interest in preserving the status quo of the global oil market price structure. One could also argue that the surge of domestic hydrocarbon production in the United States has been driven by market concerns. Major domestic policy initiatives, such as the development of the Keystone XL, have been influenced by national security concerns (in addition to creating jobs). Those security concerns revolve around the independence not simply from foreign oil states, but rather from the ability of one or more foreign actors to control the price of US energy consumption.

Market Power Motivation in Other Types of States

This discussion suggests a final question related to scope conditions that revolves around the types of states that can be motivated to take aggressive actions to secure market power for their firms. Two dimensions of state characteristics that are especially pertinent given our case selection are state-firm relations and regime type. On the first dimension, we focused our empirical analysis on market power opportunities for states that own or have significant control over their firms in a market. While limiting the scope of our case studies in this way was helpful to investigate the plausibility of our theory, it does leave open the question of whether similar dynamics can occur in states with less control over their firms. We have seen that such states may be willing to take significant steps to block other states from gaining market power, as when the United States took military action to prevent Iraq from invading Saudi Arabia and acquiring a significant share of the world's oil reserves. The motivations behind market power prevention, however, do not necessarily need to be connected to the market position of one's own firms. Instead, these actions may be aimed at protecting domestic consumers or limiting the ability of a rival state to gain political leverage from increased market power.

Determining if and when these states would take aggressive steps to provide their own firms with market power is a more complex question. States that do not own or have significant control over firms may be less able to enjoy the benefits of market power. Since they are not able to directly access rents or control prices, these states may find it more difficult to generate revenue or political leverage from their firms' price-setting abilities. Firms will thus need to find a way to motivate these states to take costly steps to increase the firms' market power. Understanding how this process might work would require more theoretical development at the domestic level. For example, firms that play a key role in the government's winning coalition may be better able to lobby states to take actions to promote their market position. Additionally, the level of state capacity may influence a state's ability to generate revenue from the taxation of the profits earned by price-setting firms. However, a more complete model of market power politics that incorporates aspects of a state's domestic political economy will have to be left to future research.

In addition to limiting our focus to a particular set of state-firm relations, our cases also only examined the effect of market power motivations on foreign policy in non-democratic states. Thus, we cannot yet speak to how these processes might differ in a democratic domestic setting. With that caveat in place, we can speculate here as to why it is difficult to imagine market power opportunities that could motivate states like the United States or states within the European Union to use war or even gray zone conflict. It is not that our theory is inapplicable to these cases—far from it. The advantage these cases have in terms of avoiding this motivation lies in the combination of their distributed involvement in multiple markets, none of which is likely to dominate the political process at any given time to the extent that leaders will have an incentive to use violence. That is not to say that these states are more peaceful than others, only that the complexity of their economies makes these market power opportunities less likely.

Moreover, there is no reason to expect that democratic states would not be able to accrue benefits from market power. While dependence upon commodity exports is generally associated with autocratic governments, some democracies also greatly benefit from resource rents.[10] Examples of democratic states that generate significant revenue from rents from commodity markets include Botswana (diamonds), Chile (copper), and Norway (hydrocarbons). However, it remains an open question as to whether democratic states would pursue aggressive foreign policies, including the use of force, in pursuit of market power. In particular, we expect that democratic leaders may face significant domestic political constraints on their behavior in addition to the international economic and institutional constraints that we have considered here. Identifying how and when these domestic constraints can influence the expansionist strategies of democracies would be a useful way to extend our model of market power politics to additional contexts.

Lessons of the Book

It is no doubt that increased economic interdependence and the development of international institutions have had many positive effects on international relations. Major interstate armed conflict is much less prevalent than in earlier

---

10. Dunning 2008; Ross 2001.

periods, and three-quarters of a century has passed since the last major power war. Institutions have emerged to provide rules and procedures that states can use to resolve property rights disputes that may have previously been settled on the battlefield. Given the strength of these institutions and the consolidation of territorial integrity norms, once these disputes are settled, they are unlikely to re-emerge. The resolution of property rights disputes reduces economic inefficiencies and may in turn increase levels of trade. This trade could then create greater levels of interdependence, which would further constrain states from using violence. Thus, the combination of economic interdependence and international institutions can potentially provide a positive feedback loop that promotes international cooperation.

However, our analysis in this book indicates that these economic and institutional processes are more complex than may initially appear. First, while economic integration can increase levels of economic interdependence between states, it can also open up potentially valuable market power opportunities. Thus, depending upon the market structure, international economic relations can both constrain states from using violence and motivate them to pursue expansionist policies. Second, while the development of strong international institutions can provide effective mechanisms for states to peacefully resolve their differences, this rule-based order can be incompatible with the goals of revisionist states.[11] Thus, some states may have incentives to eschew institutional rules and procedures and instead push their interests through extra-institutional channels. Let us take a look at each of these lessons in more detail and then consider how they help to inform our understanding of the contemporary use of gray zone tactics by countries like China and Russia.

Markets and International Conflict

A key implication of our theory of market power politics is that economic relationships can both constrain and motivate international conflict. When states are economically interdependent, they face significant costs of exiting their economic relationships. War can severely interrupt economic interactions between states. Due to high exit costs, economic interdependence can thus make military conflict more costly and risky. Given this, when a state is dependent upon another state, it can be constrained from escalating disputes militarily. In this way, as liberal theorists expect, increased economic ties can sometimes reduce the threat of war.

---

11. Indeed, this seems true almost by definition of what it means to be a revisionist state.

As we have shown, however, economic ties can also motivate states to take aggressive actions. In particular, states that want to provide their firms in hard commodity markets with price-setting capabilities may be able to achieve an increase in market power through territorial expansion. Greater global interconnectedness can increase the value of these market power opportunities to states. For one, with greater international trade, there is the potential for an increased flow of rents from exports to price-setting firms. Additionally, in interconnected markets, there will likely be more opportunities to use bargaining leverage derived from market power. Given these significant benefits, states may be willing to turn to the use of force or gray zone tactics in pursuit of market power opportunities.

These contrasting effects of economic exchange on conflict behavior result from the fact that economic competition can motivate states differently depending upon the market environment they find themselves in. In normal economic exchange environments, buyers and sellers are price takers. Without an ability to affect prices, political actors in this environment largely focus their efforts on coordinating or manipulating the effects of the market price. In such an environment, economic incentives often drive states toward cooperation, institution building, and efficiency. On the other hand, when firms have the opportunity to gain price-setting capabilities in markets of imperfect competition, competitive pressure can drive states to focus on maximizing their firms' market share. This market power competition between states, which largely mirrors the behavior of rent-seeking firms, can create multiple inefficiencies. For one, when states take costly actions to pursue market power or to prevent others from doing so, they forgo opportunities to engage in more productive activities. Additionally, if a state is successful in guaranteeing price-setting capabilities for its firms, this can lead to the establishment of inefficient market structures.

To illustrate the dynamics in these different economic environments, consider the market power competition in the natural gas market that we explored in Chapter 6. The goal for Russia is to set the price of natural gas without suffering large changes in production. To do so, Russia has to diminish its competition as it preserves its export markets in Europe. As a state with few profitable sources of revenue, Russia relies heavily on hydrocarbon exports. With the ability to set its own price for exporting natural gas to Europe without experiencing a precipitous decline in European consumption, it can leverage rents from its international gas customers to provide government revenue and subsidize domestic consumers. Additionally, this market power gives Russia leverage when negotiating with states that rely upon Russian gas. At the same time, European gas customers have incentives to take costly steps to block Russia's attempts to corner the market. Russia's consumers can be pushed too far, which would lead to their attempt to exit the market in favor of more expensive green energy technologies.

This market power competition environment differs from the economic conditions that Russia would face in a perfectly competitive market within a normal economic exchange environment. If Russia had no ability to constrain other exporting states from developing natural gas pipeline capacity to deliver gas to Europe, then its policy goals and activities would be markedly different. Instead of investing in weakening its competition, Russia would focus on efficiently delivering as much gas to its customers as possible. Thus, competitive pressures can lead states to take very different paths depending upon the structure of the market. By establishing precisely how these market structures affect the incentives of states, we can better understand the cross-cutting pressures that economic relationships have on international conflict behavior.

Limits of International Institutions

International institutions offer a peaceful avenue for states to efficiently resolve disputes over property rights. Institutions provide rules as to how states should allocate property rights, as well as procedures that can help states reach settlements in line with those rules. However, as we have seen, some of the same factors that make these institutions effective can also deter states from reaching an institutional settlement. Institutional rules provide focal points that can help states coordinate, but they also limit the set of settlements available to disputants. Similarly, reputation costs can increase the incentive of states to commit to institutional agreements, but they also raise the stakes of reaching any agreement in the first place. If states find that their political goals are not compatible with the rules laid out by international institutions, they will be reluctant to use institutional dispute-resolution mechanisms.

Thus, in the context of international property rights, the establishment of international institutions can be a double-edged sword. Stronger institutions can more effectively resolve disputes, but they can also reduce the incentive of states to use them. We saw an example of this in China's behavior in the ongoing maritime dispute over the South China Sea. Previous research indicates that arbitration is one of the most effective ways to settle an international boundary dispute. However, when the Philippines attempted to use the Permanent Court of Arbitration to resolve aspects of the disputed claim, China declined to participate and rejected the court's ruling. China's reluctance to pursue such an institutional solution is not surprising given that any potential settlement under UNCLOS would likely require China to give up its claim to much of the South China Sea. Given the incompatibility between the institutional rules and its expansionist goals, China opted to strategically delay any resolution of the dispute.

We expect that a similar dynamic can be generalized to institutions beyond the realm of property rights. International institutions provide rules that help states coordinate and cooperate more efficiently and peacefully across a wide range of issue areas. However, a rule-based order is inherently biased toward maintaining the status quo. Thus, states motivated to push for significant changes in a given issue area may find that they are unable do so within the existing institutional structures. To achieve these revisionist goals, these states will have incentives to skirt the rules and eschew institutional procedures.

This highlights the tensions that arise in establishing a rule-based order in a global system characterized by changes in power over time. The structure of international institutions generally reflects the interests of the most powerful states in the system when they were established.[12] By their nature, these institutional rules cannot always take into account shifting power dynamics. Over time, we will see the emergence of states who will want to push for changes at the global level that are inconsistent with these rules. These revisionist states will have incentives to push their interests through extra-institutional channels. Given constraints of economic interdependence, these states may find the costs of change through war too costly. Instead, they may push for changes through the use of gray zone tactics.

Such a strategy not only allows these states to achieve short-term gains, it also contributes to the erosion of existing rules and norms. In this way, revisionist states can promote incremental changes in international institutions that better reflect their interests.[13] Let us consider in more detail the lessons our research provides on the use of these gray zone tactics in contemporary global politics.

The "Gray Zone" May Be Here to Stay

We began our discussion in Chapter 1 with a description of some of the unconventional activities pursued by China and Russia to expand their territorial reach. Foreign policy observers have labeled these activities "gray zone tactics" because they fall somewhere in the gray zone between conventional diplomacy and war. As several observers have noted, gray zone activities are not a new phenomenon in international relations.[14] However, they do appear to play a central role in the current foreign policy portfolios of key global actors, including China and Russia. Moreover, this shift to gray zone activities may be a challenge for countries like the United States, whose foreign policies are largely structured to respond to

---

12. Gilpin 1981.

13. North 1990.

14. Brands 2016; Kapusta 2015; Mazarr 2015.

more conventional activities. Thus, it is important to understand the strategic dynamics underlying this trend in international relations.

Our study of market power politics provides insights into the use of gray zone activities, especially in the context of property rights disputes. Mazarr argues that gray zone campaigns are defined by three components: measured revisionism, strategic gradualism, and unconventional tools.[15] First, states utilizing these tactics want to revise the status quo, but they would prefer to do so without completely disrupting the existing international system. Second, they try to shift the status quo gradually over time through salami tactics and faits accomplis. Finally, they use unconventional tools that fall below the threshold of traditional warfare. In many ways, this corresponds to our conceptualization of strategic delay of property rights disputes in the face of market power opportunities. Through strategic delay, states have the opportunity to pursue gray zone tactics that allow them to gradually achieve their market power goals over time or increase their bargaining power in future negotiations over property rights.

The model of market power politics that we have developed in this book identifies a key factor that can motivate states to revise the status quo, as well as conditions under which they will prefer to pursue these revisionist goals in a gradual, measured capacity. While there are many reasons why states would want to expand their territorial reach or alter the status quo distribution of international property rights, we focused our attention on an understudied driver of expansionist activity: the desire to capture market power. To secure the ability to set the prices of a valuable good, states may need to gain access to additional resources or stave off market competitors. As we have discussed, states have a variety of strategies that they could use to pursue these market power goals, including violence, peaceful dispute settlement, and strategic delay. When states are economically interdependent and their revisionist goals are not compatible with the rules laid out by international institutions, they may be constrained from using violence and have incentives to avoid institutional settlements. In these situations, states may opt for a strategy of delay, which provides them with an opportunity to use gray zone tactics to further their goals.

While our study has focused on the politics surrounding the desire for market power, we expect that similar dynamics would likely play out in a wide range of international disputes. Regardless of what motivates states to pursue expansionist foreign policies, economic interdependence and international institutions will always provide potential constraints on their behavior. Thus, economically interdependent states should generally be less willing to militarily escalate

---

15. Mazarr 2015.

international property rights disputes regardless of whether there is a potential market power opportunity. Additionally, the compatibility of international institutions in a given issue area should influence when states will decide to strategically delay dispute resolution and pursue gray zone tactics to achieve their goals.

Given this, our study of market power politics indicates that the turn to the gray zone is, in part, a product of our contemporary global environment. The international system of the twenty-first century is marked by increased economic interdependence and an increasingly institutionalized rule-based order. These structures have their roots in the institutions established by the United States and its allies after World War II, and they have become even more pervasive in recent decades as more countries have become increasingly more integrated into the international economy and international institutions. As two legs of the so-called Kantian tripod, economic interdependence and international institutions are generally viewed as factors that promote cooperation and reduce the occurrence of armed conflict between countries.[16] However, as we have shown, the configuration of these factors can sometimes lead countries to pursue strategic delay rather than the peaceful settlement of disputes.

In a more economically interdependent world, we expect that states will be less willing to escalate disputes militarily. Because they face increased exit costs, interdependent states are reluctant to disrupt their economic relationships by turning to violence. This fits in with the expectations of liberal theorists. However, the pressures that have historically motivated states to pursue expansionist and aggressive tactics have not completely disappeared. If the overt use of force is taken off the table, states will look for other ways to achieve these goals. One aim of the creation of international institutions is to create rules and procedures that states can use to settle their disputes peacefully. However, such a rule-based order is generally biased toward maintaining the status quo. Thus, revisionist states may find it difficult to achieve their goals within the existing institutional structures and may instead turn to gray zone tactics.

We expect that gray zone tactics will likely continue to be the foreign policy tool of choice for states like China and Russia in the near future. As long as these states remain integrated into the globalized economy, either they or their rivals will likely be constrained from escalating to full-scale war in pursuit of their foreign policy goals. For example, many of China's neighbors in the Asia-Pacific region have some level of economic dependence on China, given its position as the world's second largest economy. Moreover, moves by China to create the Asian Infrastructure Investment Bank and implement its One Belt, One Road

16. Russett and Oneal 2001.

initiative will likely contribute to a number of countries finding themselves in an extended asymmetrical interdependent relationship with China. As for Russia, the general structure of the European gas market is not likely to change significantly, at least in the medium term. Russia's economy will likely continue to depend upon hydrocarbon sales to European markets, while many European countries will continue to need to rely upon Russian gas for a significant portion of their energy needs.

At the same time, we expect that both China and Russia are likely to remain what Mazarr calls "measured revisionists."[17] They are motivated to push for changes in the international status quo, but given their integration in the global economy, they will prefer to work within existing international structures to achieve their goals whenever possible. However, as we have seen in the previous chapters, these countries' ambitions are not always compatible with the rules laid out by international institutions. In these situations, both China and Russia have shown that they are willing to work around institutions if need be. In the South China Sea, China has ignored the unfavorable ruling of the Permanent Court of Arbitration and has continued its "cabbage strategy" of gradually expanding its area of control in the sea. Similarly, Russia has not been deterred by the lack of international recognition for its annexation of Crimea, and it continues to pursue its borderization strategy in South Ossetia. As long as these gray zone tactics remain effective means of pushing their interests, we expect countries like China and Russia to continue using them.

Policy Options

What implications do these complex relationships between markets, institutions, and conflict behavior have for policymaking at the global level? In particular, what steps could policymakers take to encourage more peaceful and efficient resolution of international property rights disputes? We see that there are two general paths available. On the one hand, policymakers could try to promote changes in economic and institutional structures to reduce the incentives of states to pursue expansionist activities or avoid institutional settlements of property rights disputes. Alternatively, they could work to identify strategies within existing structures to deter expansionist strategies of violence and delay. While both of these options have their merits, they also face some practical limitations. Thus, we expect that there will no magic bullet to end the type of pernicious activities that we have discussed in this book.

---

17. Mazarr 2015.

Fostering Market Complexity

If we take a moment to revisit the conditions shaping Russia's policy choices that were discussed in Chapter 6, what changes in the Russian economy might mitigate Putin's willingness to use gray zone tactics and risky expansionist violence? One characteristic of the Russian economy that stands out is its relative simplicity with respect to the number of markets that matter. Beyond hydrocarbons and weapons exports, Russia has a dearth of influential industries that might counter the motivation that drives its current policies. Were Russia to develop an influential manufacturing sector, for example, that also relied on exports to the European Union, an internal counterbalance to the market power pressures coming out of the hydrocarbon industry could emerge.

As of 2019, however, such a diversification looks unlikely. Intensified sanctions against Russia have significantly impacted the economy, increasing the state's dependence on the revenues coming from hydrocarbons and weapons.[18] Stepping back from this example, and looking at the diverse economies in the European Union and the United States, one can see how the complexity of market pressures makes it less likely for any one market power opportunity to dominate policymakers' decision calculus.

Redesigning Institutions

A more extreme approach to addressing the issues that inhibit peaceful dispute resolution would be to redesign international institutions. When states are reluctant to use international institutions to resolve property rights disputes, it is largely due to two factors. First, institutions constrain the set of potential settlements available to disputants. This may take some otherwise acceptable agreements off the table. Second, since boundary agreements are difficult to renegotiate once they are settled, the stakes of the settlement of a property rights dispute are especially high. Incorporating more flexibility into international institutions could potentially mitigate some of these effects and increase the willingness of actors to pursue an institutional settlement. However, this would likely sacrifice some of the benefits that institutions provide.

More flexible international rules would expand the set of potential settlements available to disputants. As we saw earlier in this chapter, the ambiguous geographic characteristics of the Caspian Sea gave the disputants leeway in

---

18. We do not wish to suggest that these sanctions are in any way inappropriate. We merely observe that their success has made the kind of market diversification required for counterpressures more difficult.

designing an agreement on its legal status. Because the Caspian Sea did not necessarily fall under the jurisdiction of UNLCOS, the littoral states did not have
to follow the rules laid out by that convention. Instead, they were able to come
to their own mutually acceptable agreement based upon the "common waters,
divided bottom" approach. Given this greater flexibility, Russia was willing to
abandon its strategy of delay and agree to settle the property rights dispute.

While adding more flexibility to international rules could open the door to
politically acceptable settlements that would otherwise be off the table, too much
flexibility would likely undermine the usefulness of institutional rules. For one,
if rules are too vague, then they will no longer provide effective focal points for
states to coordinate on. Without these focal points, negotiation over the distribution of property rights would be more complex and perhaps more time consuming. Additionally, as we have seen in the East China Sea, ambiguous rules
raise the stakes for states submitting their property rights claims to arbitration or
adjudication, given the uncertainty about what ruling will be made. Thus, while
flexible rules make states more willing to pursue negotiated settlements in line
with institutional rules, they may decrease their willingness to use legal dispute-
resolution procedures.

One could also potentially incorporate flexibility into institutional arrangements to account for shifts in power over time. This could reduce the stakes of
accepting institutional settlements to property rights disputes. Scholars have
found that flexible provisions like escape clauses and finite time horizons can
make states more willing to accept institutional agreements. For example, the
inclusion of escape clauses makes international trade agreements easier to reach
when leaders are uncertain about future domestic demands.[19] Similarly, in issue
areas where states are more uncertain about future shocks, they are more likely to
sign renegotiable agreements with finite time horizons.[20]

While the incorporation of such flexibility provisions may be beneficial in
some issue areas, like trade and investment treaties, we expect that they are a less
realistic solution for property rights disputes. For one, a key benefit of settling
territorial and maritime disputes is that they clarify property rights for economic
actors. These flexibility conditions would increase the possibility for future shifts
in these property rights. Such uncertainty could have negative effects on trade
and investment. Perhaps more importantly, allowing for the widespread use of
escape clauses and renegotiation in property rights settlements would require
abandoning the norm of territorial integrity that has developed over recent

---

19. Rosendorff and Milner 2001.

20. Koremenos 2005.

decades. This would increase the international legitimacy of pressing territorial claims and likely spark new property rights disputes.

Deterring Gray Zone Tactics

Rather than attempting to overhaul international institutions, one could instead accept the limitations of these institutions and find ways to work within them. By recognizing the economic and institutional factors that lead states to pursue inefficient strategies of violence or strategic delay, policymakers can try to identify strategies to deter expansionist behavior when these motivations arise. In essence, this option accepts the reality of strategic delay and instead focuses on anticipating potential scenarios where strategic delay and gray zone expansionism may occur.

What policy options might counter these outcomes? Looking back at the early years of the Cold War, we see a similar debate unfolding regarding the salami tactics of Soviet expansionism. The policy solution then was to raise the costs for the Soviet Union should it decide to slice off more territory. A central component of the Truman Doctrine was to place American interests (and lives) firmly in the way of the path of expansion. This option is not without risk, of course, and raises the stakes of any potential conflict. Credibility issues become paramount in these scenarios, as was the case throughout the Cold War.

Concluding Thoughts

These potential policy options have commonalities that are worth highlighting, as they illustrate the essential message of the book. Both the market diversification and institutional reform solutions aim to ameliorate the underlying conditions that create the market power opportunities in the first place, while the last solution accepts these conditions as they are and raises costs to make war and gray zone tactics less likely. All three of these solutions accept the core dynamics of the market power politics in play.

These dynamics have a powerful underlying effect on international politics when market power opportunities emerge, and policy solutions that ignore them will be much more likely to fail or unravel quickly. If our international institutions do not provide nonviolent options to resolve border disputes that account for and ameliorate the motivations that emerge with market power, they will continue to be ignored by states that cannot afford to pass up these opportunities. No amount of institutional reform will help if it does not offer solutions to this set of political-economic interactions.

Similarly, policy responses outside of international institutions will find durable solutions to be evasive. The set of sanctions imposed by the West on Russia in response to its expansion into Ukraine and the annexation of Crimea can only provide a temporary deterrent to Russia given the current market power opportunity it faces and the lack of domestic counter-pressures. When policies such as these sanctions appear to be merely stalling the target, rather than opening up a dialogue to resolve the issues at stake, an underlying market power mechanism may be the root cause.

Thus, our work suggests that states such as China and Russia will be unlikely to accept or adhere to political solutions to their territorial and maritime disputes until the underlying market power issues are resolved. At the same time, as long as neither state can freely manipulate its markets without fear of adaptation by its major partners, this ceiling-constraint on its market power will act as friction and reduce the likelihood of war. As long as green energy remains a more expensive alternative for western European states than Russian natural gas, Russia will be able to leverage that power, but it cannot push so hard that it pushes western European states into going fully green at any expense. Similarly, as long as advanced manufacturing states such as Japan and the United States can buy the rare earths they need from China, China is likely to maintain its market power in that arena. If China were to use its stranglehold on the rare earths market as a tool of economic statecraft, however, both Japan and the United States would relaunch their domestic programs and reduce China's market power.

Of course, like any exploratory work, this book raises more questions than it answers. But our goal of highlighting the unraveling effects of market power opportunities provides us with a fresh insight into some of the most important conflicts in modern history. The Persian Gulf War was a pivotal moment in American foreign policy, solidifying the United States' pole position as the last remaining superpower. But at the same time, it launched a new era of transnational terrorism that would go on to consume vast resources around the globe. Fears of German market power (particularly with respect to coal) fueled the Allied resistance to German expansion during the first half of the twentieth century. And in the seventeenth century, competition over beaver pelt markets drove intense and durable violence among the indigenous tribes in the northeast region of North America. These events may not be frequent (thankfully), but they have had great impact on the world, and it behooves us to continue studying the underlying political-economy dynamics that can make normally pacific economic ties a motivation for war.

Bibliography

Abbott, Kenneth W., and Duncan Snidal. 2000. Hard and Soft Law in International Governance. *International Organization* 54 (3): 421–456.

Abdelal, Rawi. 2013. The Profits of Power: Commerce and Realpolitik in Eurasia. *Review of International Political Economy* 20 (3): 421–456.

Admati, Anat R., and Motty Perry. 1987. Strategic Delay in Bargaining. *Review of Economic Studies* 54 (3): 345–364.

Alibo, Dia Sotto, and Yoshiyuki Nozaki. 2000. Dissolved Rare Earth Elements in the South China Sea: Geochemical Characterization of the Water Masses. *Journal of Geophysical Research: Oceans* 105 (C12): 28771–28783.

Allee, Todd L., and Paul K. Huth. 2006. Legitimizing Dispute Settlement: International Legal Rulings as Domestic Political Cover. *American Political Science Review* 100 (2): 219–234.

Allison, Roy. 2008. Russia Resurgent? Moscow's Campaign to "Coerce Georgia to Peace." *International Affairs* 84 (6): 1145–1171.

Almoguera, Pedro A., Christopher C. Douglas, and Ana María Herrera. 2011. Testing for the Cartel in OPEC: Non-Cooperative Collusion or Just Non-Cooperative? *Oxford Review of Economic Policy* 27 (1): 144–168.

Altman, Dan. 2017. By Fait Accompli, Not Coercion: How States Wrest Territory from Their Adversaries. *International Studies Quarterly* 61 (4): 881–891.

Altman, Dan. 2018. Advancing without Attacking: The Strategic Game around the Use of Force. *Security Studies* 27 (1): 58–88.

Anscombe, Frederick F. 1997. *The Ottoman Gulf: The Creation of Kuwait, Saudi Arabia, and Qatar.* New York: Columbia University Press.

Asariotis, Regina, and Anila Premti. 2018. Conservation and Sustainable Use of Marine Biodiversity in Areas beyond National Jurisdiction: Recent Legal Developments. *UNCTAD Transport and Trade Facilitation Newsletter* 79. Available at <https://unctad.org/fr/pages/newsdetails.aspx?OriginalVersionID=1861>.

Asia Maritime Transparency Initiative. 2015. Diplomacy Changes, Construction Continues: New Images of Mischief and Subi Reefs. June 18. Available at <https://amti.csis.org/diplomacy-changes-construction-continues-new-images-of-mischief-and-subi-reefs/>.

Asia Maritime Transparency Initiative. 2018. Comparing Aerial and Satellite Images of China's Spratly Outposts. February 16. Available at <https://amti.csis.org/comparing-aerial-satellite-images-chinas-spratly-outposts/>.

Asia Maritime Transparency Initiative. 2019a. Signaling Sovereignty: Chinese Patrols at Contested Reefs. September 26. Available at <https://amti.csis.org/signaling-sovereignty-chinese-patrols-at-contested-reefs/>.

Asia Maritime Transparency Initiative. 2019b. Update: China Risks Flare-Up over Malaysian, Vietnamese Gas Resources. December 13. Available at <https://amti.csis.org/china-risks-flare-up-over-malaysian-vietnamese-gas-resources/>.

Astakhova, Oleysa, and Can Selzer. 2020. Turkey, Russia Launch TurkStream Pipeline Carrying Gas to Europe. *Reuters*. January 8. Available at <https://www.reuters.com/article/us-turkey-russia-pipeline/turkey-russia-launch-turkstream-pipeline-carrying-gas-to-europe-idUSKBN1Z71WP>.

Atzili, Boaz. 2011. *Good Fences, Bad Neighbors: Border Fixity and International Conflict*. Chicago: University of Chicago Press.

Baldwin, David. 1980. Interdependence and Power: A Conceptual Analysis. *International Organization* 34 (4): 471–506.

Barbieri, Katherine. 1995. Economic Interdependence and Militarized Interstate Conflict, 1870–1985. PhD diss., State University of New York, Binghamton.

Barbieri, Katherine. 1996. Economic Interdependence: A Path to Peace or a Source of Interstate Conflict? *Journal of Peace Research* 33 (1): 29–49.

Basedau, Matthias, and Jann Lay. 2009. Resource Curse or Rentier Peace? The Ambiguous Effects of Oil Wealth and Oil Dependence on Violent Conflict. *Journal of Peace Research* 46 (6): 757–776.

Bautista, Lowell, and Clive Schofield. 2012. Philippine-China Border Relations: Cautious Engagement amid Tensions. In *Beijing's Power and China's Borders: Twenty Neighbors in Asia*, edited by Bruce A. Elleman, Stephen Kotkin, and Clive Schofield, 235–250. Armonk, NY: M. E. Sharpe.

BBC News. 2009. UN Defines Romania-Ukraine Border. February 3. Available at <http://news.bbc.co.uk/2/hi/europe/7867683.stm>.

BBC News. 2010. Boat Collisions Spark Japan-China Diplomatic Row. September 8. Available at <https://www.bbc.com/news/world-asia-pacific-11225522>.

Beblawi, Hazem, and Giacomo Luciani, eds. 1987. *The Rentier State*. London: Croom Helm.

Belyi, Andrei. 2015. *Transnational Gas Markets and Euro-Russian Energy Relations*. Basingstoke, UK: Palgrave Macmillan.

Bilsborough, Shane. 2012. The Strategic Implications of China's Rare Earths Policy. *Journal of Strategic Security* 5 (3): 1–12.

Blockmans, Steven. 2015. Crimea and the Quest for Energy and Military Hegemony in the Black Sea Region: Governance Gap in a Contested Geostrategic Zone. *Journal of Southeast European & Black Sea Studies* 15 (2): 179–189.

Boban, Davor, and Karlo Lončar. 2016. Geopolitical Consequences of Resolving the Legal Status of the Caspian Sea: Security and Energy Aspects. *Hrvatski Geografski Glasnik* 78 (2): 77–100.

Booth, William, and Will Englund. 2014. Gunmen's Seizure of Parliament Building Stokes Tensions in Ukraine's Crimea. *Washington Post*. February 27. Available at <https://www.washingtonpost.com/world/crimean-city-offers-refuge-to-police-who-battled-protesters-in-ukraine-capital/2014/02/26/ad105a78-9ec3-11e3-9ba6-800d1192d08b_story.html>.

BP. 2010. *BP Statistical Review of World Energy 2010*. Available at <http://www.dartmouth.edu/~cushman/courses/engs41/BP-StatisticalReview-2010.pdf>.

BP. 2018. *BP Statistical Review of World Energy 2018*. Available at <https://www.bp.com/content/dam/bp/business-sites/en/global/corporate/pdfs/energy-economics/statistical-review/bp-stats-review-2018-full-report.pdf>.

BP. 2019. *BP Statistical Review of World Energy 2019*. Available at <https://www.bp.com/content/dam/bp/business-sites/en/global/corporate/pdfs/energy-economics/statistical-review/bp-stats-review-2019-full-report.pdf>.

Bradsher, Keith. 2019. China's Supply of Minerals for iPhones and Missiles Could Be a Risky Trade Weapon. *New York Times*. May 23. Available at <https://www.nytimes.com/2019/05/23/business/china-us-trade-war-rare-earths.html>.

Brands, Hal. 2016. Paradoxes of the Gray Zone. *Foreign Policy Research Institute*. February 5. Available at <https://www.fpri.org/article/2016/02/paradoxes-gray-zone/>.

Brecher, Michael, and Jonathan Wilkenfeld. 1997. *A Study of Crisis*. Ann Arbor: University of Michigan Press.

Bruce, Chloë. 2007. Power Resources: The Political Agenda in Russo-Moldovan Gas Relations. *Problems of Post-Communism* 54 (3): 29–47.

Brutschin, Elina. 2015. Shaping the EU's Energy Policy Agenda: The Role of Eastern European Countries. In *Energy Policy Making in the EU: Building the Agenda*, edited by Jale Tosun, Sophie Biesenbender, and Kai Schulze, 187–204. London: Springer.

Bueno de Mesquita, Bruce, and Randolph Siverson. 1995. War and the Survival of Political Leaders: A Comparative Study of Regime Types and Political Accountability. *American Political Science Review* 89 (4): 841–855.

Burke, Edmund, Timothy Heath, Jeffrey Hornung, Logan Ma, Lyle Morris, and Michael Chase. 2018. *China's Military Activities in the East China Sea: Implications for Japan's Air Self-Defense Force*. Santa Monica, CA: RAND.

Bush, President George H. W. 1990. *Address to the Nation Announcing the Deployment of United States Armed Forces to Saudi Arabia.* August 8. Available at <https://bush-41library.tamu.edu/archives/public-papers/2147>.

Butler, Charles J. 2014. Rare Earth Elements: China's Monopoly and Implications for US National Security. *The Fletcher Forum for World Affairs* 38 (1): 23–39. Available at <http://www.fletcherforum.org/wp-content/uploads/2014/04/38-1_Butler1.pdf>.

Cao, Hong, Zhi-lei Sun, Chang-ling Liu, En-tao Liu, Xue-jun Jiang, and Wei Huang. 2018. Origin of Natural Sulfur-Metal Chimney in the Tangyin Hydrothermal Field, Okinawa Trough: Constraints from Rare Earth Element and Sulfur Isotopic Compositions. *China Geology* 1 (2): 225–235.

Carter, David B., and H. E. Goemans. 2011. The Making of the Territorial Order: New Borders and the Emergence of Interstate Conflict. *International Organization* 65 (2): 275–309.

Carter, David B., and H. E. Goemans. 2014. The Temporal Dynamics of New International Borders. *Conflict Management and Peace Science* 31 (3): 285–302.

Casey, Michael S. 2007. *The History of Kuwait.* Westport, CT: Greenwood Press.

Cede, Frank. 2009. The Settlement of International Disputes by Legal Means: Arbitration and Judicial Settlement. In *The SAGE Handbook of Conflict Resolution,* edited by Jacob Bercovitch, Victor Kremenyuk, and I. William Zartman, 358–375. Los Angeles: Sage.

Chatagnier, J. Tyson, and Kerim Can Kavaklı. 2015. From Economic Competition to Military Combat. *Journal of Conflict Resolution* 61 (7): 1510–1536.

Chaudoin, Stephen. 2014. Promises or Policies? An Experimental Analysis of International Agreements and Audience Reactions. *International Organization* 68 (1): 235–256.

Chivers, C. J. 2006. Explosions in Southern Russia Sever Gas Lines to Georgia. *New York Times.* January 23. Available at <https://www.nytimes.com/2006/01/23/world/europe/explosions-in-southern-russia-sever-gas-lines-to-georgia.html>.

Coffey, Luke. 2018. *NATO Membership for Georgia: In U.S. and European Interest.* Heritage Foundation Special Report No. 199. Available at <https://www.heritage.org/sites/default/files/2018-01/SR-199_0.pdf>.

Copeland, Dale. 1996. Economic Interdependence and War: A Theory of Trade Expectations. *International Security* 20 (4): 5–41.

Cormac, Rory, and Richard J. Aldrich. 2018. Grey Is the New Black: Covert Action and Implausible Deniability. *International Affairs* 94 (3): 477–494.

Cox, Clinten, and Jindrich Kynicky. 2018. The Rapid Evolution of Speculative Investment in the REE Market before, during, and after the Rare Earth Crisis of 2010–2012. *Extractive Industries and Society* 5 (1): 8–17.

Cramton, Peter C. 1992. Strategic Delay in Bargaining with Two-Sided Uncertainty. *Review of Economic Studies* 59 (1): 205.

Crescenzi, Mark J.C. 2005. *Economic Interdependence and Conflict in World Politics.* Lanham, MD: Lexington Books.

Crescenzi, Mark J.C. 2018. *Of Friends and Foes: Reputation and Learning in International Politics.* New York: Oxford University Press.

Crystal, Jill. 1995. *Oil and Politics in the Gulf: Rulers and Merchants in Kuwait and Qatar.* Cambridge: Cambridge University Press.

Dahlman, Carl J. 1979. The Problem of Externality. *Journal of Law and Economics* 22 (1): 141–162.

Dai, Xinyuan. 2007. *International Institutions and National Policies.* Cambridge: Cambridge University Press.

Dawisha, Adeed. 2009. *Iraq: A Political History from Independence to Occupation.* Princeton, NJ: Princeton University Press.

Dickel, Ralf, Elham Hassanzadeh, James Henderson, Anouk Honoré, Laura El-Katiri, Simon Pirani, Howard Rogers, Jonathan Stern, and Katja Yafimava. 2014. *Reducing European Dependence on Russian Gas: Distinguishing Natural Gas Security from Geopolitics.* Oxford Institute for Energy Studies Paper NG 92. Available at <https://doi.org/10.26889/9781784670146>.

Duelfer, Charles. 2004. *Comprehensive Report of the Special Advisor to the DCI on Iraq's WMD.* Volume 1. Washington, DC: Central Intelligence Agency.

Duffield, John. 2007. What Are International Institutions? *International Studies Review* 9 (1): 1–22.

Dunning, Thad. 2008. *Crude Democracy: Natural Resource Wealth and Political Regimes.* Cambridge: Cambridge University Press.

Dupuy, Florian, and Pierre-Marie Dupuy. 2013. A Legal Analysis of China's Historic Rights Claim in the South China Sea. *American Journal of International Law* 107 (1): 124–141.

Eckstein, Harry. 1975. Case Study and Theory in Political Science. In *Handbook of Political Science*, Vol. 7: *Strategies of Inquiry*, edited by Fred I. Greenstein and Nelson W. Polsby, 79–138. Reading, MA: Addison-Wesley.

El Ghoneimy, Mohd. Talaat. 1966. The Legal Status of the Saudi-Kuwaiti Neutral Zone. *The International and Comparative Law Quarterly* 15 (3): 690–717.

Ellickson, Bryan. 1993. *Competitive Equilibrium: Theory and Applications.* New York: Cambridge University Press.

European Union. 2009. Directive 2009/73/EC of the European Parliament and of the Council of 13 July 2009 Concerning Common Rules for the Internal Market in Natural Gas and Tepealing Directive 2003/55/EC. *Official Journal of the European Union* L 211: 94–132.

Fearon, James D. 1996. Bargaining over Objects That Influence Future Bargaining Power. Paper presented at the Annual Meeting of the American Political Science Association, Washington, DC, August, 28–31.

Fearon, James D. 1998. Bargaining, Enforcement, and International Cooperation. *International Organization* 52 (2): 269–305.

Felgenhauer, Pavel. 2009. After August 7: The Escalation of the Russia-Georgia War. In *The Guns of August 2008: Russia's War in Georgia*, edited by Svante E. Cornell and S. Frederick Starr, 143–161. Armonk, NY: M. E. Sharpe.

Finnie, David H. 1992. *Shifting Lines in the Sand: Kuwait's Elusive Frontier with Iraq*. Cambridge, MA: Harvard University Press.

Fravel, M. Taylor. 2008. *Strong Borders, Secure Nation: Cooperation and Conflict in China's Territorial Disputes*. Princeton, NJ: Princeton University Press.

Fravel, M. Taylor. 2011. China's Strategy in the South China Sea. *Contemporary Southeast Asia* 33 (3): 292–319.

Friedman, Thomas L. 1990. The Iraqi Invasion: Battle of the Saudi Soil. *New York Times*. August 4. Available at <https://www.nytimes.com/1990/08/04/world/the-iraqi-invasion-battle-for-the-saudi-soil.html>.

Ganguli, Rajive, and Douglas R. Cook. 2018. Review Rare Earths: A Review of the Landscape. *MRS Energy & Sustainability* 5: E9.

Garfinkel, Michelle R., and Stergios Skaperdas. 2000. Conflict Without Misperceptions or Incomplete Information: How the Future Matters. *Journal of Conflict Resolution* 44 (6): 793–807.

Gause, F. Gregory, III. 2002. Iraq's Decisions to Go to War, 1980 and 1990. *Middle East Journal* 56 (1): 47–70.

Gent, Stephen E., and Megan Shannon. 2010. The Effectiveness of International Arbitration and Adjudication: Getting into a Bind. *Journal of Politics* 72 (2): 366–380.

Gent, Stephen E., and Megan Shannon. 2011. Decision Control and the Pursuit of Binding Conflict Management: Choosing the Ties That Bind. *Journal of Conflict Resolution* 55 (5): 710–734.

Gent, Stephen E., and Megan Shannon. 2014. Bargaining Power and the Arbitration and Adjudication of Territorial Claims. *Conflict Management and Peace Science* 31 (3): 303–322.

Gilpin, Robert. 1981. *War and Change in World Politics*. Cambridge: Cambridge University Press.

Gleditsch, Kristian S. 2002. Expanded Trade and GDP Data. *Journal of Conflict Resolution* 46 (5): 712–724.

Gowa, Joanne. 1995. Democratic States and International Disputes. *International Organization* 49 (3): 511–522.

Grant, John P., and J. Craig Barker. 2009. *Parry and Grant Encylopaedic Dictionary of International Law*. 3rd ed. Oxford: Oxford University Press.

Grant, Thomas D. 2017. Frozen Conflicts and International Law. *Cornell International Law Journal* 50: 361–413.

Green, Michael, Kathleen Hicks, Zack Cooper, John Schaus, and Jake Douglas. 2017. *Countering Coercion in Maritime Asia: The Theory and Practice of Gray Zone Deterrence*. Washington, DC: Center for Strategic and International Studies.

Greenwood, Christopher. 1991. Iraq's Invasion of Kuwait: Some Legal Issues. *Cambridge Review of International Affairs* 5 (1): 21–31.

Grigas, Agnia. 2017. *The New Geopolitics of Natural Gas*. Cambridge, MA: Harvard University Press.

Gunay, Niyazi. 2000. *Arab League Summit Conferences, 1964–2000*. Washington Institute Policywatch 496. Available at <https://www.washingtoninstitute.org/policy-analysis/view/arab-league-summit-conferences-19642000>.

Guzman, Andrew T. 2008. *How International Law Works: A Rational Choice Theory*. New York: Oxford University Press.

Harper, Jo. 2017. Nordstream II Gas Pipeline in Deep Water. *DW*. November 14. Available at <http://www.dw.com/en/nordstream-ii-gas-pipeline-in-deep-water/a-41372833>.

Hawley, T. M. 1992. *Against the Fires of Hell: The Environmental Disaster of the Gulf War*. New York: Harcourt Brace Jovanovich.

Hayton, Bill. 2018. The Modern Creation of China's "Historic Rights" Claim in the South China Sea. *Asian Affairs* 49 (3): 370–382.

Hearty, Grace, and Mayaz Alam. 2019. Rare Earths: Next Element in the Trade War? *Center for Strategic and International Studies*. August 20. Available at <https://www.csis.org/analysis/rare-earths-next-element-trade-war>.

Heather, Patrick. 2015. *The Evolution of European Traded Gas Hubs*. Oxford Institute for Energy Studies Paper NG 104. Available at <https://doi.org/10.26889/9781784670467>.

Heather, Patrick, and Beatrice Petrovich. 2017. *European Traded Gas Hubs: An Updated Analysis on Liquidity, Maturity and Barriers to Market Integration*. Oxford Institute for Energy Studies Energy Insight: 13. Available at <https://doi.org/10.26889/eii3.201705>.

Hedlund, Stefan. 2014. *Putin's Energy Agenda: The Contradictions of Russia's Resource Wealth*. Boulder, CO: Lynne Rienner.

Hirschman, Albert O. 1945. *National Power and the Structure of Foreign Trade*. Berkeley: University of California Press.

Högselius, Per. 2013. *Red Gas: Russia and the Origins of European Energy Dependence*. New York: Palgrave Macmillan.

Holsti, Kalevi J. 1991. *Peace and War: Armed Conflicts and International Order, 1648–1989*. Cambridge: Cambridge University Press.

Hongo, Yayoi, Hajime Obata, Toshitaka Gamo, Miwako Nakaseama, Junichiro Ishibashi, Uta Konno, Shunsuke Saegusa, Satoru Ohkubo, and Urumu Tsunogai. 2007. Rare Earth Elements in the Hydrothermal System at Okinawa Trough Back-Arc Basin. *Geochemical Journal* 41 (1): 1–15.

Huth, Paul K., Sarah E. Croco, and Benjamin J. Appel. 2011. Does International Law Promote the Peaceful Settlement of International Disputes? Evidence from the Study of Territorial Conflicts since 1945. *American Political Science Review* 105 (2): 415–436.

Ibrahim, Youssef. 1990. Iraq Said to Prevail in Oil Dispute with Kuwait and Arab Emirates. *New York Times*. July 26. Available at <https://www.nytimes.com/1990/07/26/world/iraq-said-to-prevail-in-oil-dispute-with-kuwait-and-arab-emirates.html>.

Illarionov, Andrei. 2009. The Russian Leadership's Preparation for War, 1999–2008. In *The Guns of August 2008: Russia's War in Georgia*, edited by Svante E. Cornell and S. Frederick Starr, 49–84. Armonk, NY: M. E. Sharpe.

International Seabed Authority. 2006. *Polymetallic Nodules*. Available at <http://www.isa.org.jm/files/documents/EN/Brochures/ENG7.pdf>.

International Seabed Authority. 2019. ISBA/25/A/15: Decision of the Assembly of the International Seabed Authority Relating to the Implementation of the Strategic Plan for the Authority for the Period 2019–2023. Available at <https://www.isa.org.jm/document/isba25a15-decision-assembly>.

Ivan, Ruxandra. 2007. Patterns of Cooperation and Conflict: Romanian-Ukrainian Bilateral Relations, 1992–2006. *Studia Europaea* 52 (2): 99–130.

Jaffe, Amy Myers. 2006. *Iraq's Oil Sector: Issues and Opportunities*. James A. Baker III Institute for Public Policy, Rice University. Available at <https://www.bakerinstitute.org/media/files/Research/e1b042f0/Iraq_s_Oil_Sector.pdf>.

Johnston, Alastair Iain. 2013. How New and Assertive Is China's New Assertiveness? *International Security* 37 (4): 7–48.

Kandiyoti, Rafael. 2015. *Powering Europe: Russia, Ukraine, and the Energy Squeeze*. New York: Palgrave Macmillan.

Kantai, Tallash, Mari Luomi, Kate Neville, and Nicole Schabus. 2020. Summary of the Twenty-sixth Annual Session of the International Seabed Authority (First Part): 17–21 February 2020. *Earth Negotiations Bulletin* 25 (224): 1–13. Available at <https://enb.iisd.org/download/pdf/enb25224e.pdf>.

Kapusta, Philip. 2015. The Gray Zone. *Special Warfare* 28 (4): 18–25.

Karsh, Efraim, and Inari Rautsi. 1991. Why Saddam Hussein Invaded Kuwait. *Survival* 33 (1): 18–30.

Keohane, Robert. 1984. *After Hegemony: Cooperation and Discord in the World Political Economy*. Princeton, NJ: Princeton University Press.

King & Spalding LLP. 2018. *LNG in Europe 2018: An Overview of Import Terminals in Europe*. Available at <https://www.kslaw.com/attachments/000/006/010/original/LNG_in_Europe_2018_-_An_Overview_of_LNG_Import_Terminals_in_Europe.pdf>.

Klinger, Julie Michelle. 2018. Rare Earth Elements: Development, Sustainability and Policy Issues. *Extractive Industries and Society* 5 (1): 1–7.

Knight, Jack, and James Johnson. 2011. *The Priority of Democracy: Political Consequences of Pragmatism*. Princeton, NJ: Princeton University Press.

Koo, Min Gyo. 2009. The Senkaku/Diaoyu dispute and Sino-Japanese Political-Economic Relations: Cold Politics and Hot Economics? *Pacific Review* 22 (2): 205–232.

Koremenos, Barbara. 2005. Contracting around International Uncertainty. *American Political Science Review* 99 (4): 549–565.

Koremenos, Barbara, Charles Lipson, and Duncan Snidal. 2001. The Rational Design of International Institutions. *International Organization* 55 (4): 761–799.

Kramer, Andrew E. 2018a. In a Prize for Big Oil Firms, Caspian Deal Eases Access. *New York Times*. October 8. Available at <https://www.nytimes.com/2018/10/08/business/energy-environment/caspian-sea-deal-eases-access.html>.

Kramer, Andrew E. 2018b. Russia and 4 Other Nations Settle Decades-Long Dispute over Caspian Sea. *New York Times*. August 12. Available at <https://www.nytimes.com/2018/08/12/world/europe/caspian-sea-russia-iran.html>.

Krueger, Anne O. 1974. The Political Economy Seeking of the Rent-Seeking Society. *American Economic Review* 64 (3): 291–303.

Kruglashov, Anatolii. 2011. Troublesome Neighborhood: Romania and Ukraine Relationship. *New Ukraine* 11: 114–124.

Kyodo. 2019. Japan Protests as China Continues Gas Activity in Contested Waters. *South China Morning Post*. February 7. Available at <https://www.scmp.com/news/china/diplomacy/article/2185272/japan-protests-china-continues-gas-activity-contested-waters>.

Landes, William M. 1971. An Economic Analysis of the Courts. *Journal of Law and Economics* 14 (1): 61–107.

Lanoszka, Alexander. 2016. Russian Hybrid Warfare and Extended Deterrence in Eastern Europe. *International Affairs* 92 (1): 175–195.

Lathrop, Coalter G. 2009. Maritime Delimitation in the Black Sea (Romania v. Ukraine). *American Journal of International Law* 103 (3): 543–549.

Latiff, Rozanna, and A. Ananthalakshmi. 2020. Malaysian Oil Exploration Vessel Leaves South China Sea Waters after Standoff. *Reuters*. May 12. Available at <https://www.reuters.com/article/us-china-security-malaysia/malaysian-oil-exploration-vessel-leaves-south-china-sea-waters-after-standoff-idUSKBN22O1M9>.

Latorre, María C., and Nobuhiro Hosoe. 2016. The Role of Japanese FDI in China. *Journal of Policy Modeling* 38 (2): 226–241.

Levy, Jack S. 2008. Case Studies: Types, Designs, and Logics of Inference. *Conflict Management and Peace Science* 25 (1): 1–18.

Lipson, Charles. 1991. Why Are Some International Agreements Informal? *International Organization* 45 (4): 495–538.

Long, Drake. 2020. "Deep Plowing the South China Sea": China Eyes Resources, Steps up Research. *Radio Free Asia*. April 7. Available at <https://www.rfa.org/english/news/china/southchinasea-research-04072020184044.html>.

Mabro, Robert. 1992. OPEC and the Price of Oil. *The Energy Journal* 13: 1–17.

MacFarquhar, Neil. 2014. Ukraine Deal Imposes Truce Putin Devised. *New York Times*. September 5. Available at <https://www.nytimes.com/2014/09/06/world/europe/ukraine-cease-fire.html>.

MacFarquhar, Neil. 2015. Ukraine's Latest Peace Plan Inspires Hope and Doubts. *New York Times*. February 12. Available at <https://www.nytimes.com/2015/02/13/world/europe/ukraine-talks-cease-fire.html>.

MacFarquhar, Neil. 2016. Warming Relations in Person, Putin and Erdogan Revive Pipeline Deal. *New York Times*. October 10. Available at <https://www.nytimes.com/2016/10/11/world/europe/turkey-russia-vladimir-putin-recep-tayyip-erdogan.html>.

Machacek, Erika, and Niels Fold. 2018. Competitors Contained? Manufacturer Strategies in the Global Rare Earth Value Chain: Insights from the Magnet Filament. *Extractive Industries and Society* 5 (1): 18–27.

Macias, Amanda. 2018. China Quietly Installed Defensive Missile Systems on Strategic Spratly Islands in Hotly Contested South China Sea. *CNBC*. May 2. Available at <https://www.cnbc.com/2018/05/02/china-added-missile-systems-on-spratly-islands-in-south-china-sea.html>.

Manzini, Paola, and Marco Mariotti. 2001. Perfect Equilibria in a Model of Bargaining with Arbitration. *Games and Economic Behavior* 37 (1): 170–195.

Martin, Lisa L., and Beth A. Simmons. 2013. International Organizations and Institutions. In *Handbook of International Relations*, edited by Walter Carlsnaes, Thomas Risse, and Beth A. Simmons, 192–211. London: Sage.

Mazarr, Michael J. 2015. *Mastering the Gray Zone: Understanding a Changing Era of Conflict*. Carlisle, PA: Strategic Studies Institute and US Army War College Press.

McGillivray, Fiona, and Alastair Smith. 2008. *Punishing the Prince: A Theory of Interstate Relations, Political Institutions, and Leader Change*. Princeton, NJ: Princeton University Press.

Melin, Molly M., and Alexandru Grigorescu. 2014. Connecting the Dots: Dispute Resolution and Escalation in a World of Entangled Territorial Claims. *Journal of Conflict Resolution* 58 (6): 1085–1109.

Mitchell, Sara McLaughlin, and Paul R. Hensel. 2007. International Institutions and Compliance with Agreements. *American Journal of Political Science* 51 (4): 721–737.

Mitrova, Tatiana. 2015. Changing Gas Price Mechanisms in Europe and Russia's Gas Pricing Policy. *IAEE Energy Forum* 24 (Antalya Special Issue): 39–40.

Morrow, James D. 1999. How Could Trade Affect Conflict? *Journal of Peace Research* 36 (4): 481–489.

Multiple Wire Reports. 1990. Iraqi Troops near Saudi Border. *St. Petersburg Times*. August 4, A1.

Myers, Steven Lee, and Ellen Barry. 2014. Putin Reclaims Crimea for Russia and Bitterly Denounces the West. *New York Times*. March 18. Available at <https://www.nytimes.com/2014/03/19/world/europe/ukraine.html>.

Nilsson, Niklas. 2018. *Russian Hybrid Tactics in Georgia*. Silk Road Paper, Central Asia-Caucasus Institute & Silk Road Studies Program. Available at <http://www.

silkroadstudies.org/publications/silkroad-papers-and-monographs/item/13274-russian-hybrid-tactics-in-georgia.html>.

Nord Stream. 2018. Nord Stream Reaches Average Utilisation of 93% in 2017–51 Bcm Delivered to the European Union. January 16. Available at <https://www.nord-stream.com/press-info/press-releases/nord-stream-reaches-average-utilisation-of-93-in-2017-51-bcm-delivered-to-the-european-union-500/>.

North, Andrew. 2015. Russian Expansion: "I Went to Bed in Georgia—and Woke Up in South Ossetia." *The Guardian*. May 20. Available at <https://www.theguardian.com/world/2015/may/20/russian-expansion-georgia-south-ossetia>.

North, Douglass C. 1990. *Institutions, Institutional Change and Economic Performance*. Cambridge: Cambridge University Press.

Nygren, Bertil. 2008. Putin's Use of Natural Gas to Reintegrate the CIS Region. *Problems of Post-Communism* 55 (4): 3–15.

O'Rourke, Ronald. 2018. *Maritime Territorial and Exclusive Economic Zone (EEZ) Disputes Involving China: Issues for Congress*. Congressional Research Service R42784.

Olearchyk, Roman. 2015. Shell to Withdraw from Shale Gas Exploration in Eastern Ukraine. *Financial Times*. June 11. Available at <https://www.ft.com/content/0c66011e-104a-11e5-bd70-00144feabdc0>.

Oliphant, Roland. 2014. "Polite People" Leading the Silent Invasion of the Crimea. *The Telegraph*. March 2. Available at <https://www.telegraph.co.uk/news/worldnews/europe/ukraine/10670547/Ukraine-crisis-Polite-people-leading-the-silent-invasion-of-the-Crimea.html>.

Ostrom, Elinor. 1990. *Governing the Commons: The Evolution of Institutions for Collective Action*. Cambridge: Cambridge University Press.

Ottaway, David B. 1996. Been There, Done That. *Washington Post*. July 21. Available at <https://www.washingtonpost.com/archive/lifestyle/1996/07/21/been-there-done-that/922cdbd4-9805-4d4e-a6ca-128723ab9f4a/>.

Paton Walsh, Nick. 2006. Georgian Leaders Turn up the Heat on Russia as Gas Supplies Are Restored after Freezing Week. *Guardian*. January 29. Available at <https://www.theguardian.com/world/2006/jan/30/georgia.oil>.

Permanent Mission of the People's Republic of China to the United Nations. 2009. Note Verbale CML/17/2009. Available at <https://www.un.org/Depts/los/clcs_new/submissions_files/mysvnm33_09/chn_2009re_mys_vnm_e.pdf>.

Philippine Statistics Authority. 2020. *The Foreign Investments in the Philippines, Fourth Quarter 2019*. Available at <https://psa.gov.ph/sites/default/files/Fourth Quarter 2019 Foreign Investments Report.pdf>.

Pillai, R. V., and Mahendra Kumar. 1962. The Political and Legal Status of Kuwait. *International and Comparative Law Quarterly* 11 (1): 108–130.

Pinchuk, Denis, Olesya Astakhova, and Oleg Vukmanovic. 2015. Gazprom to Offer More Gas at Spot Prices via Nord Stream II. *Reuters*. October 13. Available at

<https://www.reuters.com/article/us-russia-gazprom-spot/exclusive-gazprom-to-offer-more-gas-at-spot-prices-via-nord-stream-ii-idUSKCN0S71XS20151013>.

Pinfari, Marco. 2009. *Nothing but Failure? The Arab League and the Gulf Cooperation Council as Mediators in Middle Eastern Conflicts.* Crisis States Research Center Working Paper, Series 2, No. 45. Available at <http://www.lse.ac.uk/international-development/Assets/Documents/PDFs/csrc-working-papers-phase-two/wp45.2-nothing-but-failure.pdf>.

Polachek, Solomon. 1980. Conflict and Trade. *Journal of Conflict Resolution* 24 (1): 55–578.

Polityuk, Pavel. 2012. Exxon, Shell-Led Group Win $10 Billion Ukraine Gas Project. *Reuters.* August 15. Available at <https://www.reuters.com/article/us-shell-exxonmobil-ukraine/exxon-shell-led-group-win-10-billion-ukraine-gas-project-idUSBRE87E0C320120815>.

Popjanevski, Johanna. 2009. From Sukhumi to Tskinvali: The Path to War in Georgia. In *The Guns of August 2008: Russia's War in Georgia,* edited by Svante E. Cornell and S. Frederick Starr, 143–161. Armonk, NY: M. E. Sharpe.

Posner, Richard A. 1975. The Social Costs of Monopoly and Regulation. *Journal of Political Economy* 83 (4): 807–827.

Powell, Emilia Justyna, and Krista E. Wiegand. 2014. Strategic Selection: Political and Legal Mechanisms of Territorial Dispute Resolution. *Journal of Peace Research* 51 (3): 361–374.

Powell, Robert. 1999. *In the Shadow of Power: States and Strategies in International Politics.* Princeton, NJ: Princeton University Press.

Powell, Robert. 2006. War as a Commitment Problem. *International Organization* 60 (1): 169–203.

President of Russia. 2018. Convention on the Legal Status of the Caspian Sea. August 12. Available at <http://en.kremlin.ru/supplement/5328>.

Prorok, Alyssa K., and Paul K. Huth. 2015. International Law and the Consolidation of Peace Following Territorial Changes. *Journal of Politics* 77 (1): 161–174.

Ramos-Mrosovsky, Carlos. 2008. International Law's Unhelpful Role in the Senkaku Islands. *University of Pennsylvania Journal of International Law* 29 (4): 903–946.

Reed, Lucy, and Kenneth Wong. 2016. Marine Entitlements in the South China Sea: The Arbitration between the Philippines and China. *American Journal of International Law* 110 (4): 746–760.

Reed, Stanley, and Sebnem Arsu. 2015. Russia Presses Ahead with Plan for Gas Pipeline to Turkey. *New York Times.* Available at <https://www.nytimes.com/2015/01/22/business/international/russia-presses-ahead-with-plan-for-gas-pipeline-to-turkey.html>.

Reed, Stanley, and Milan Schreuer. 2018. E.U. Settles with Russia's Gazprom over Antitrust Charges. *New York Times.* May 24. Available at <https://www.nytimes.com/2018/05/24/business/energy-environment/eu-gas-gazprom.html>.

Reuters. 2012. Ukraine Picks Shell, Chevron to Develop Shale Gas Fields. *Reuters*. May 11. Available at <https://www.reuters.com/article/shell-chevron-ukraine/ukraine-picks-shell-chevron-to-develop-shale-gas-fields-idUSL5E8GBAE020120511>.

Reuters. 2013. Ukraine Signs Oil, Gas Deal with Eni and EDF, Sees $4 Billion Investment. *Reuters*. November 17. Available at <https://www.reuters.com/article/us-ukraine-energy-deal-idUSBRE9AQ0JX20131127>.

Reuters. 2014a. Exxon Says Pursuit of Ukraine Skifska Block on Hold. *Reuters*. March 5. Available at <https://www.reuters.com/article/exxon-outlook-ukraine/exxon-says-pursuit-of-ukraine-skifska-block-on-hold-idUSWEN00CO620140305>.

Reuters. 2014b. Shell Pulled Out of Gas Field Talks in Ukraine in January. *Reuters*. March 19. Available at <https://www.reuters.com/article/ukraine-crisis-shell/shell-pulled-out-of-gas-field-talks-in-ukraine-in-january-idUSL6N0MG40H20140319>.

Reuters. 2014c. Ukraine Says Chevron Plans to Pull Out of $10 Bln Shale Gas Deal. *Reuters*. December 15. Available at <https://www.reuters.com/article/ukraine-crisis-gas/ukraine-says-chevron-plans-to-pull-out-of-10-bln-shale-gas-deal-idUSL6N0TZ29A20141215>.

Reuters. 2018. U.S. Condemns Syria's Ties with Georgian Breakaway Regions. *Reuters*. May 30. Available at <https://www.reuters.com/article/us-georgia-syria-usa/u-s-condemns-syrias-ties-with-georgian-breakaway-regions-idUSKCN1IV1GS>.

Ridgewell, Henry. 2018. Europe Split on Nord Stream 2 Pipeline as US Warns against Dependence on Russian Gas. *VOA News*. March 8. Available at <https://www.voanews.com/europe/europe-split-nord-stream-2-pipeline-us-warns-against-dependence-russian-gas>.

Roach, J. Ashley. 2014. *Malaysia and Brunei: An Analysis of Their Claims in the South China Sea*. CNA Occasional Paper. Available at <https://www.cna.org/CNA_files/PDF/IOP-2014-U-008434.pdf>.

Rosendorff, B. Peter, and Helen V. Milner. 2001. The Optimal Design of International Trade Institutions: Uncertainty and Escape. *International Organization* 55 (4): 829–857.

Ross, Michael L. 2001. Does Oil Hinder Democracy? *World Politics* 53 (3): 325–361.

Russett, Bruce M., and John R. Oneal. 2001. *Triangulating Peace: Democracy, Interdependence, and International Organizations*. New York: Norton.

Sanger, David E., and Rick Gladstone. 2015. Piling Sand in a Disputed Sea, China Literally Gains Ground. *New York Times*. April 8. Available at <https://www.nytimes.com/2015/04/09/world/asia/new-images-show-china-literally-gaining-ground-in-south-china-sea.html>.

Schelling, Thomas C. 1960. *The Strategy of Conflict*. Cambridge, MA: Harvard University Press.

Schelling, Thomas. 1966. *Arms and Influence*. New Haven, CT: Yale University Press.

Schultz, Kenneth A. 2014. What's in a Claim? De Jure versus De Facto Borders in Interstate Territorial Disputes. *Journal of Conflict Resolution* 58 (6): 1059–1084.

Schultz, Kenneth A. 2015. Borders, Conflict, and Trade. *Annual Review of Political Science* 18 (1): 125–145.

Scott, S. D. 2001. Deep Ocean Mining. *Geoscience Canada* 28 (2): 87–96.

Seaman, John. 2010. *Rare Earths and Clean Energy: Analyzing China's Upper Hand.* Notes de l'Ifri, French Institute of International Relations. Available at <https://www.ifri.org/en/publications/enotes/notes-de-lifri/rare-earths-and-clean-energy-analyzing-chinas-upper-hand>.

Seaman, John. 2019. *Rare Earths and China: A Review of Changing Criticality in the New Economy.* Notes de l'Ifri, French Institute of International Relations. Available at <https://www.ifri.org/en/publications/notes-de-lifri/rare-earths-and-china-review-changing-criticality-new-economy>.

Shaw, Han-yi. 1999. The Diaoyutai/Senkaku Islands Dispute: Its History and an Analysis of the Ownership Claims of the P.R.C., R.O.C., and Japan. *Maryland Series in Contemporary Asian Studies* 1999 (3): Article 1.

Shen, Lei, Na Wu, Shuai Zhong, and Li Gao. 2017. Overview on China's Rare Earth Industry Restructuring and Regulation Reforms. *Journal of Resources and Ecology* 8 (3): 213–222.

Shen, Yuzhou, Ruthann Moomy, and Roderick G. Eggert. 2020. China's Public Policies Toward Rare Earths, 1975–2018. *Mineral Economics.* Available at <https://doi.org/10.1007/s13563-019-00214-2>.

Shenon, Philip. 1995. Manila Sees China Threat on Coral Reef. *New York Times.* February 19. Available at <https://www.nytimes.com/1995/02/19/world/manila-sees-china-threat-on-coral-reef.html>.

Shevchenko, Vitaly. 2014. "Little Green Men" or "Russian Invaders"? *BBC.* March 11. Available at <https://www.bbc.com/news/world-europe-26532154>.

Simmons, Beth A. 2000a. International Law and State Behavior: Commitment and Compliance in International Monetary Affairs. *American Political Science Review* 94 (4): 819–835.

Simmons, Beth A. 2000b. The Legalization of International Monetary Affairs. *International Organization* 54 (3): 573–602.

Simmons, Beth A. 2002. Capacity, Commitment, and Compliance: International Law and the Settlement of Territorial Disputes. *Journal of Conflict Resolution* 46 (6): 829–856.

Simmons, Beth A. 2005. Rules over Real Estate: Trade, Territorial Conflict, and International Borders as Institutions. *Journal of Conflict Resolution* 49 (6): 823–848.

Skaperdas, Stergios. 1992. Cooperation, Conflict, and Power in the Absence of Property Rights. *American Economic Review* 82 (4): 720–739.

Smith, Jeffrey J. 2010. Brunei and Malaysia Resolve Outstanding Maritime Boundary Issues. *LOS Reports* 1: 1–4. Available at <https://ssrn.com/abstract=2296908>.

Smith, Nicholas Ross. 2016. *EU-Russian Relations and the Ukraine Crisis.* Cheltenham, UK: Edward Elgar.

Smith, Sheila A. 2012. Japan and the East China Sea Dispute. *Orbis* 56 (3): 370–390.

Smith Stegen, Karen. 2011. Deconstructing the "Energy Weapon": Russia's Threat to Europe as Case Study. *Energy Policy* 39 (10): 6505–6513.

State Council, People's Republic of China. 2015. "Made in China 2025" Plan Issued. Available at <http://english.www.gov.cn/policies/latest_releases/2015/05/19/content_281475110703534.htm>.

Storey, Ian James. 1999. Creeping Assertiveness: China, the Philippines and the South China Sea Dispute. *Contemporary Southeast Asia* 21 (1): 95–118.

Stytus, Andrius. 2017. Lithuania Receives First LNG from the United States. *Reuters*. August 21. Available at <https://www.reuters.com/article/us-lithuania-lng/lithuania-receives-first-lng-from-the-united-states-idUSKCN1B11BW>.

Suman, D O. 1981. A Comparison of the Law of the Sea Claims of Mexico and Brazil. *Ocean Development & International Law* 10 (1): 131–173.

Tarar, Ahmer. 2016. A Strategic Logic of the Military Fait Accompli. *International Studies Quarterly* 60 (4): 742–752.

Terzian, Pierre. 1991. The Gulf Crisis: The Oil Factor: An Interview with Pierre Terzian. *Journal of Palestinian Studies* 20 (2): 100–105.

Thibaut, John, and Laurens Walker. 1978. A Theory of Procedure. *California Law Review* 66 (3): 541–566.

Thies, Cameron, and Timothy Peterson. 2015. *Intra-Industry Trade: Cooperation and Conflict in the Global Political Economy*. Stanford, CA: Stanford University Press.

Tollison, Robert D. 1982. Rent Seeking: A Survey. *Kyklos* 35 (4): 575–602.

Tsai, Chung-min. 2016. Sino-Japanese Relations over the East China Sea: The Case of Oil and Gas Fields. *Journal of Territorial and Maritime Studies* 3 (2): 71–87.

Tullock, Gordon. 1967. The Welfare Costs of Tarrifs, Monopolies, and Theft. *Economic Inquiry* 5 (3): 224–232.

United Nations. 1982. United Nations Convention on the Law of the Sea. Available at <https://www.un.org/Depts/los/convention_agreements/texts/unclos/UNCLOS-TOC.htm>.

United Nations Security Council. 2020. Resolutions Adopted by the Security Council in 1990. Available at <https://www.un.org/securitycouncil/content/resolutions-adopted-security-council-1990>.

United States Department of Defense. 2017. *Annual Report to Congress: Military and Security Developments Involving the People's Republic of China 2017*. Available at <https://dod.defense.gov/Portals/1/Documents/pubs/2017_China_Military_Power_Report.PDF>.

US Energy Information Association. 2013a. *Caspian Sea Region*. Available at <https://www.eia.gov/international/content/analysis/regions_of_interest/Caspian_Sea/caspian_sea.pdf>.

US Energy Information Association. 2013b. *Technically Recoverable Shale Oil and Shale Gas Resources: An Assessment of 137 Shale Formations in 41 Countries Outside the*

United States. Available at <https://www.eia.gov/analysis/studies/worldshalegas/pdf/overview.pdf>.

US Energy Information Association. 2014. *East China Sea Tensions*. Available at <https://www.eia.gov/beta/international/regions-topics.cfm?RegionTopicID=ECS>.

US Energy Information Association. 2019. Oil and Petroleum Products Explained. Available at <https://www.eia.gov/energyexplained/oil-and-petroleum-products/>.

Valencia, Mark J. 2007. The East China Sea Dispute: Context, Claims, Issues, and Possible Solutions. *Asian Perspective* 31 (1): 127–167.

Vasquez, John A. 1993. *The War Puzzle*. Cambridge: Cambridge University Press.

Vavilov, Andrey, ed. 2015a. *Gazprom: An Energy Giant and Its Challenges in Europe*. Basingstoke, UK: Palgrave Macmillan.

Vavilov, Andrey. 2015b. Introduction. In *Gazprom: An Energy Giant and Its Challenges in Europe*, edited by Andrey Vavilov, 1–14. Basingstoke, UK: Palgrave Macmillan.

Vavilov, Andrey, and Georgy Trofimov. 2015a. European Challenges: Competitive Pressure, Gas-Market Liberalization, and the Crisis of Long-Term Contracting. In *Gazprom: An Energy Giant and Its Challenges in Europe*, 139–179. Basingstoke, UK: Palgrave Macmillan.

Vavilov, Andrey, and Georgy Trofimov. 2015b. The Struggle for Pipelines: Gazprom's Attempts at Strategic Expansion in the "Near Abroad." In *Gazprom: An Energy Giant and Its Challenges in Europe*, 105–138. Basingstoke, UK: Palgrave Macmillan.

Vekasi, Kristin. 2014. China's Political Rise and Japan's Economic Risk: Multinational Corporations and Political Uncertainty. PhD diss., University of Wisconsin-Madison.

Verleger, Philip K., Jr. 1990. Understanding the 1990 Oil Crisis. *The Energy Journal* 11: 15–33.

Volvo. 2018. *Supply Chain Sustainability Management*. Available at <https://assets.volvocars.com/~/media/ccs/suppliers/supply-chain-sustainability-management-2018.pdf>.

Vu, Khanh. 2020. Vietnam Protests Beijing's Sinking of South China Sea Boat. *Reuters*. April 4. Available at <https://www.reuters.com/article/us-vietnam-china-southchinasea/vietnam-protests-beijings-sinking-of-south-china-sea-boat-idUSKBN21M072>.

Wagner, R. Harrison. 1988. Economic Interdependence, Bargaining Power, and Political Influence. *International Organization* 42 (3): 461–483.

Walter, Barbara F. 2009. *Reputation and Civil War: Why Separatist Conflicts Are So Violent*. Cambridge: Cambridge University Press.

Whaley, Floyd. 2012. Philippines and China in a Standoff at Sea. *New York Times*. April 12. Available at <https://www.nytimes.com/2012/04/12/world/asia/diplomatic-resolution-sought-in-south-china-sea-standoff.html>.

Wheedon, Alan. 2019. Warfare's Next Frontier Might Be the Ocean Floor—Here's Why. *ABC News (Australia)*. Available at <https://www.abc.net.au/news/2019-10-19/the-seabed-might-be-warfares-next-frontier/11606522>.

Wiegand, Krista E. 2009. China's Strategy in the Senkaku/Diaoyu Islands Dispute: Issue Linkage and Coercive Diplomacy. *Asian Security* 5 (2): 170–193.

Wiegand, Krista E. 2011. *Enduring Territorial Disputes: Strategies of Bargaining, Coercive Diplomacy, and Settlement*. Athens: University of Georgia Press.

Williams, James L. 1999. OPEC: Market Share within OPEC. *WTRG Economics*. Available at <http://wtrg.com/opecshare.html>.

Williamson, Oliver E. 1975. *Markets and Hierarchies: Analysis and Antitrust Implications*. New York: Free Press.

Williamson, Oliver E. 1985. *The Economic Institutions of Capitalism: Firms, Markets, Relational Contracting*. New York: Free Press.

Williamson, Oliver E. 1996. *The Mechanisms of Governance*. New York: Oxford University Press.

Witthoeft, Andrew. 2015. The Heavy-Handed Russian Move Nobody's Talking About. *The Diplomat*. August 6. Available at <https://thediplomat.com/2015/08/the-heavy-handed-russian-move-nobodys-talking-about/>.

World Bank. 2013. *China 2030: Building a Modern, Harmonious, and Creative Society*. Available at <http://documents.worldbank.org/curated/en/781101468239669951/China-2030-building-a-modern-harmonious-and-creative-society>.

World Bank. 2019. *Russia Economic Report #42: Weaker Global Outlook Sharpens Focus on Domestic Reforms*. Available at <http://documents.worldbank.org/curated/en/290511577724289579/pdf/Weaker-Global-Outlook-Sharpens-Focus-on-Domestic-Reforms.pdf>.

World Bank. 2020. World Integrated Trade Solution (WITS). Available at <https://wits.worldbank.org/>.

Wright, Joseph, Erica Frantz, and Barbara Geddes. 2013. Oil and Autocratic Regime Survival. *British Journal of Political Science* 45 (2): 287–306.

Xu, Fangjian, Anchun Li, Tiegang Li, Kehui Xu, Shiyue Chen, Longwei Qiu, and Yingchang Cao. 2011. Rare Earth Element Geochemistry in the Inner Shelf of the East China Sea and Its Implication to Sediment Provenances. *Journal of Rare Earths* 29 (7): 702–709.

Zacher, Mark W. 2001. The Territorial Integrity Norm: International Boundaries and the Use of Force. *International Organization* 55 (2): 215–250.

Zha, Daojiong, and Mark J. Valencia. 2001. Mischief Reef: Geopolitics and Implications. *Journal of Contemporary Asia* 31 (1): 86–103.

Zhang, Ketian. 2019. Cautious Bully: Reputation, Resolve, and Beijing's Use of Coercion in the South China Sea. *International Security* 44 (1): 117–159.

Zhang, Qiling, Zengqian Hou, and Shaohua Tang. 2010. Organic Composition of Sulphide Ores in the Okinawa Trough and Its Implications. *Acta Geologica Sinica-English Edition* 75 (2): 196–203.

Zimnitskaya, Hanna, and James von Geldern. 2011. Is the Caspian Sea a Sea; And Why Does It Matter? *Journal of Eurasian Studies* 2 (1): 1–14.

Index